KB040042

Sam Choi 선생님의

DIGITAL SAT MATH 만점공략 특강

Sam Choi 선생님의

Digital SAT Math 만점공략 특강

발 행 2023년 10월 05일 초판 1쇄
2024년 08월 30일 초판 2쇄
저 자 최승권
발행인 최영민
발행처 헤르몬하우스
주 소 경기도 파주시 신촌로 16
전 화 031-8071-0088
팩 스 031-942-8688
전자우편 hermonh@naver.com
출판등록 2015년 3월 27일
등록번호 제406-2015-31호

ISBN 979-11-92520-62-9 (53410)

Sam Choi 선생님의

DIGITAL SAT MATH 만점공략 특강

SAT Math 800점 만점을 위한 최적의 내용 & 구성

최승권(Sam Choi) 지음

Preface

안녕하세요. 국제학교 수학 전문 강사 Sam Choi입니다.

Digital SAT Math에서 반드시 만점을 받기를 원하시나요? 이 책을 선택하여 공부해야겠다고 결정하였다면 탁월한 선택이라고 자신합니다. 사실 Digital SAT Math에서 다루는 내용은 어려운 난이도의 이론을 공부해야 하는 것도 아니고 고난도 문항들로 이루어져 있지도 않습니다. 국제학교 학년을 기준으로 8~9학년의 수학 수준만으로 Math Section 만점 획득이 가능한 시험입니다. 그런데 과연 정말로 만점 획득이 쉽다고 할 수 있을까요?

소위 수학 좀 한다고 생각하는 학생들도 정작 SAT의 Math Section의 점수에서 발목을 잡혀 SAT 1500점 이상의 고득점을 획득하지 못하는 사례가 빈번함을 알고 계시리라 생각합니다. 사실 SAT의 Math Section의 만점 방법은 간단합니다. 시험에서 제시된 모든 문제를 맞히면 됩니다. 그런데 그다지 어려운 난이도의 문제들로 구성되어 있지 않은 SAT의 Math Section의 모든 문항을 실수 없이 모두 맞추는 일은 결코 쉬운 일이 아닙니다.

그 이유는 무엇일까요?

첫째, Digital SAT Math에서 다루는 수학 개념 공부를 소홀히 하였기 때문입니다. 더 엄밀하게 이야기하면 Digital SAT Math에서 다루는 개념들은 쉽다고 자만하고 기본 수학 이론 학습을 스킵하고 바로 실전 연습을 하는 경우가 가장 흔한 이유라고 할 수 있습니다. Digital SAT는 College Board에서 출제하는 Standard Test이므로 시험에 나오는 주제와 토픽들이 명확합니다. 이를 꼭 기억해두고 관련 개념들을 철저하게 공부해야 합니다.

둘째, Digital SAT Math Section에서만 출제가 되는 문제 유형을 체계적으로 정리하지 못하고 시험을 치르는 경우입니다. 소위 Digital SAT Math Section만의 간단히 숨겨진 전형적인 함정 문제들의 유형을 공부하지 않고 시험을 치른 경우입니다.

셋째, 시간 관리형 실전연습을 제대로 하지 않고 시험을 치르는 경우입니다. Digital SAT Math Section의 만점 획득의 기본 중의 기본은 주어진 시간 안에 모든 문항을 정확하고 빠르게 두 번을 풀어내며 본연의 실수 문항을 잡아내는 것입니다.

마지막으로 Digital SAT Math Section에서 항상 틀리는 그 한, 두 문항을 제대로 정리하지 않고 시험을 치르는 경우입니다. 학생마다 각자 자신이 틀리는 문제 유형이 있고, 그 문제 유형은 시간이 지나서 다시 풀어봐도 다시 틀릴 확률이 매우 높습니다. 즉 오답 문제의 정리, 오답 정리 노트는 만점 획득의 필수 요소입니다.

이 책의 특징

Digital SAT Math 만점공략특강의 특징은 이렇습니다.

1. Digital SAT Math에서 출제되는 모든 수학 개념과 이론을 충실히 담았습니다.
2. 그리고 그 수학 개념들을 바로 연습해 볼 수 있는 핵심 문항들을 엄선하여 고르고 골라 쉬운 문제부터 어려운 문제들로 바로 연습해볼 수 있도록 구성하였습니다.
3. 이 책에 담긴 모든 문제는 개념과 이론 공부를 탄탄히 하고 필수적으로 풀어봐야 하는 핵심 문제와 수많은 학생이 틀려온 오답 문제들을 모아 체계적으로 정리했습니다.
4. 누구나 다 맞추는 쉬운 문제보다는 어려운 문제들이 많이 포함되어 있습니다. 총 200문제로 구성된 핵심 문제와 오답 문제들의 구성은 쉽지만, 반드시 풀어봐야 하는 문제와 학생 대부분이 어려워하는 문제들을 정리한 것이므로 실제 Digital SAT Math Section의 시험의 문제 난이도보다 체감 난이도가 더 어렵게 구성되었습니다.
5. Final Practice Test 두 세트를 담았습니다. 모든 이론을 정리한 후 전형적인 Practice Test가 아닌 함정 문제와 난이도 있는 문제들을 다수 포함한 Final Practice Test를 통해 마지막까지 Digital SAT Math에서 꼭 만점을 받기를 희망하는 학생들의 약점을 발견하고 보완하도록 Test 세트를 구성하였습니다.
6. 계산기 Ti 84, Ti -Nspire CAS를 처음 사용하는 학생들을 위한 Digital SAT Math에서 필요한 계산기 필수 사용법을 정리하였습니다. 계산기가 필요한 실전 문제들도 풀어보면서 계산기 사용법도 함께 학습할 수 있습니다.
7. 마지막으로 본 책을 더욱 충실히 공부하기를 원하는 학생들을 위하여 저자 직강을 유학생 전문으로 인터넷 강의를 제공하는 최고의 사이트인 마스터프렙(www.masterprep.net)을 통해 마련해 두었으므로 인강을 통해 이 책을 함께 학습한다면 Digital SAT Math 만점 획득에 더욱 가까이 다가갈 수 있다고 확신합니다.

학습 과정이 충실하다면 반드시 결과를 얻을 수밖에 없습니다. 이 책을 통해 Digital SAT를 준비하는 모든 학생이 Math Section에서 반드시 만점 획득을 얻기를 희망합니다.

끝으로 Sam Choi 선생님을 믿고 함께 공부하고 있고 공부해온 많은 제자들과 지지해주시는 학부모님들께 진심으로 감사드립니다. 해외에서 십수 년을 함께 하고 있는 북경샘스아카데미 동료들에게도 감사의 인사를 드립니다. 그리고 마스터프렙의 권주근 대표님께 감사의 마음을 드립니다. 또한 진실된 교육의 동반자로 그리고 선배로서 조언해주시고 늘 응원해 주시는 네이버 재외국민특례카페 백명주 선생님과 장성훈 선배님, 그리고 특별한 파트너 김동혁 선생님께도 감사의 말씀을 드립니다.

누구보다도 사랑하는 소중한 가족들에게도 감사의 메시지를 남깁니다.

저자 Sam Choi

저자소개

국제학교 수학 전문 강사 Sam Choi 선생님은,

2007년부터 현재까지 해외에서 수많은 국제학교 학생을 지도하며, 전 세계 주요 대학의 입학과 입시실적 및 GPA 성적의 향상, 수학 관련 시험에서 만점 획득을 끌어내며 오랜 기간 오직 학업 결과를 통해 자신의 교수법을 입증하고 실력을 인정받고 있습니다.

또한 SAT Math, IB Math, AP Calculus 및 Statistics, ACT Math, AMC 등의 분야에서 영미권 수학 전문가로서 국제학교 학생들과 학부모님들의 높은 지지를 받고 있습니다.

Mr. Sam Choi's Personal History Short Cuts :

고려대학교(서울본교) 졸업-수학, 생명 유전공학 복수 전공

현) 마스터프렙 Math Subject 대표강사

현) 북경 샘스아카데미(Sam's Academy)에서 SAT 및 IB, AP, A-level 등의 국제학교 커리큘럼 기반의 수학 과목 전반을 강의 중

전) 중국 북경 수인 리벳 사립 국제학교 등의 학교에서 국제부/고등부 수학 교사로 재직하였음

해외에서 수십 년간 국제학교 학생 대상 SAT Math, AP Calculus 및 AP Statistics 관련 수많은 만점자 배출, IB Maths Subject FHL, HL, SL, Studies SL, Revised MAA HL/SL, MAI HL/SL 지도 (Since 2007 to Now) 만점자 배출 (학생들의 수강 후기는 www.ibsam.net 에서 확인할 수 있습니다.)

저서로는 〈IB Math 만점공략특강 시리즈〉 / 〈New SAT Math Essentials for Full Marks〉 / 〈수학 공부의 기술〉 / 〈고등수학의 핵심 공식집 시리즈〉 등 다수가 있습니다.

차 례

차례

☑ The Structure of Digital SAT

The digital SAT is composed of two sections: Reading and Writing and Math. Students have 64 minutes to complete the Reading and Writing section and 70 minutes to complete the Math section for a total of 2 hours and 14 minutes.

Each section is divided into 2 equal length modules, and there is a 10-minute break between the Reading and Writing section and the Math section. The first module of each section contains a broad mix of easy, medium, and hard questions. Based on how students perform on the first module, the second module of questions will either be more difficult or less difficult.

Component	Time Allotted (minutes)	Number of Questions/Tasks
Reading and Writing	64 (two 32-minute modules)	54
Math	70 (two 35-minute modules)	44
Total	134	98

☑ Paper SAT vs New Digital SAT

Features of Test	Paper SAT	Digital SAT
Test Format	SAT is a paper-based test.	Digital SAT is a computer-based test.
Exam Sections	It has 3 Sections consisting of Reading, Writing, and Math Sections.	It is a bisectional test consisting of a Reading & Writing Section and a Math Section.
Test Duration	The test duration is 3 hours and 15 minutes.	The test duration is 2 hours and 14 minutes.
Test Adaptivity	The paper-based SAT did not have this adaptive feature.	The Digital SAT will be two-stage adaptive. So based on the first module of each section, the difficulty of questions in the second module will be determined.
Length of In-context Questions	The average length of in-context questions in SAT was longer than the new mode of examination.	The average length of in-context questions (commonly known as word problems) has been reduced in digital SAT. This change has been made keeping in mind the application of Math skills in the academic and real world.
Length of Passages	SAT had longer reading passages. So limited topics were covered in this mode of test. Also, there were multiple questions asked for a single passage.	The Digital SAT Reading and Writing section will have several short passages instead of long passages. These passages will cover a wide range of topics that will determine the college readiness of the students. Furthermore, every short passage will have a single question.
Use of Calculator	Calculators were allowed for SAT Math section but there were separately timed no-calculator and calculator-allowed sections. Allowance of the calculator was restricted in SAT.	Calculators are allowed during the course of the entire Math section. However, test takers should abide by the calculator policy if they want to bring their calculator, or they can use the graphing calculator that is built into the testing application.

Features of Test	Paper SAT	Digital SAT
Test Security	The paper and pencil test didn't have the adaptive test feature, so every test taker had to give the same test which increased the probability of sharing answers. Hence SAT was less secure than Digital SAT.	Digital SAT tests will be more secure. Since it will be a two-stage adaptive test, every test taker will get a unique form of test so sharing answers is impossible in this mode.
Test Attempts	The SAT test used to happen five times a year in all international locations including India.	The Digital SAT will be held seven times a year in all international locations including India. Test takers get two extra attempts in Digital SAT.
Score Scale	The scoring scale is 1600.	The Digital SAT will continue to be scored on the 400-1600 scale.
Fees	The SAT Test fee is $109.	The Digital SAT Test fee is $109.
Results	SAT results used to take a few weeks.	Digital SAT results will be faster, and the test taker will receive their results within a few days.

☑ Paper SAT Math vs New Digital SAT Math

SAT Math	
Paper SAT Math	**New Digital SAT Math**
Time: 80min	Time: 70min
Format :	Calculator allowed for all questions
One No calculator section 15 multiple questions +5 student-produced response questions	44 Math questions
One calculator section 30 multiple questions +8 student-produced response questions	33 multiple-choice questions and around 11 student-produced response, or grid-in, questions.
KEY FACT: On the new SAT, about 75% of Math questions are multiple-choice, and about 25% are grid-in.	

☑ The Math Section Overview : 4 Math Topic Categories

Let's look at the 4 broad topic categories that Math questions on the digital test fall into. In the new SAT Math section, you'll see questions that fall into the following 4 broad categories:

- Algebra (13-15 questions): concepts such as linear equations, linear functions, and linear inequalities
- Advanced Math (13-15 questions): concepts such as nonlinear equations, nonlinear functions, and equivalent expressions
- Problem-Solving and Data Analysis (5-7 questions): ratios, rates, percentages, probability, scatterplots, statistics, and more
- Geometry and Trigonometry (5-7 questions): area and volume, right triangles, circles, and more

☑ High–level breakdown of each concept

■ Algebra:

This is your meat–and–potatoes algebra, the basic stuff. No exponents next to your 'x's. This is what we call linear equations. $4x+1=7$. Of course, the test won't ask you to solve basic equations like that. Instead, it'll give you really long word problems in which the solutions amounts to something like $3n-3=12$. And assuming that's the right equation, all you'll have to do is solve for 'n'.

■ Advanced Math:

This is the part most are dreading, high–order polynomials. However, often it's nothing more than the ax^2+bx+c variety will often be buried under a mass of verbiage, as in a 12–line word problem that you must solving using a polynomial. Often, you'll have to find creative ways to balance the equation and solve for 'x'.

■ Problem Solving and Data Analysis:

This is basically the graph and table section: bar charts, pie graphs, tedious tables with a bunch of figures for you to sort through. There will also be a fair number of word problems that ask anything involving ratios and percents, to questions dealing with median and mode.

■ Additional Topics:

This is the frustratingly vague section in which all the remainders got thrown in to, in no particular order, they are geometry, coordinate geometry and trigonometry.

☑ How to get a high score in The Digital SAT math

The changes to the Digital SAT math are designed to make it more similar to the tests you take in math class, meaning you'll be asked harder questions in a more straightforward way.

■ Heavily focused on algebra

As I mentioned above, one of the goals of the new SAT is to make it more similar to what you do in school and what you'll need for college. One part of this realignment is shifting the focus of the test towards algebra. 61% of the questions will deal with algebra topics, including manipulating equations and expressions, writing equations to solve word problems, solving quadratics, and working with formulas.

■ More data analysis

The proportion of questions focused on data analysis is also increasing. Almost a third of the questions on the test will deal with manipulating ratios and percents and understanding graphs and charts.

■ Very little geometry

With so much of the new SAT math devoted to algebra and data analysis, there is very little room for geometry. In fact, only about 6 questions will ask about geometry and trigonometry, though the test still provides most of the common formulas you'll need.

■ Still has grid-ins

Like the current version, the redesigned math section includes Student-Produced Response questions, commonly known as grid-ins.

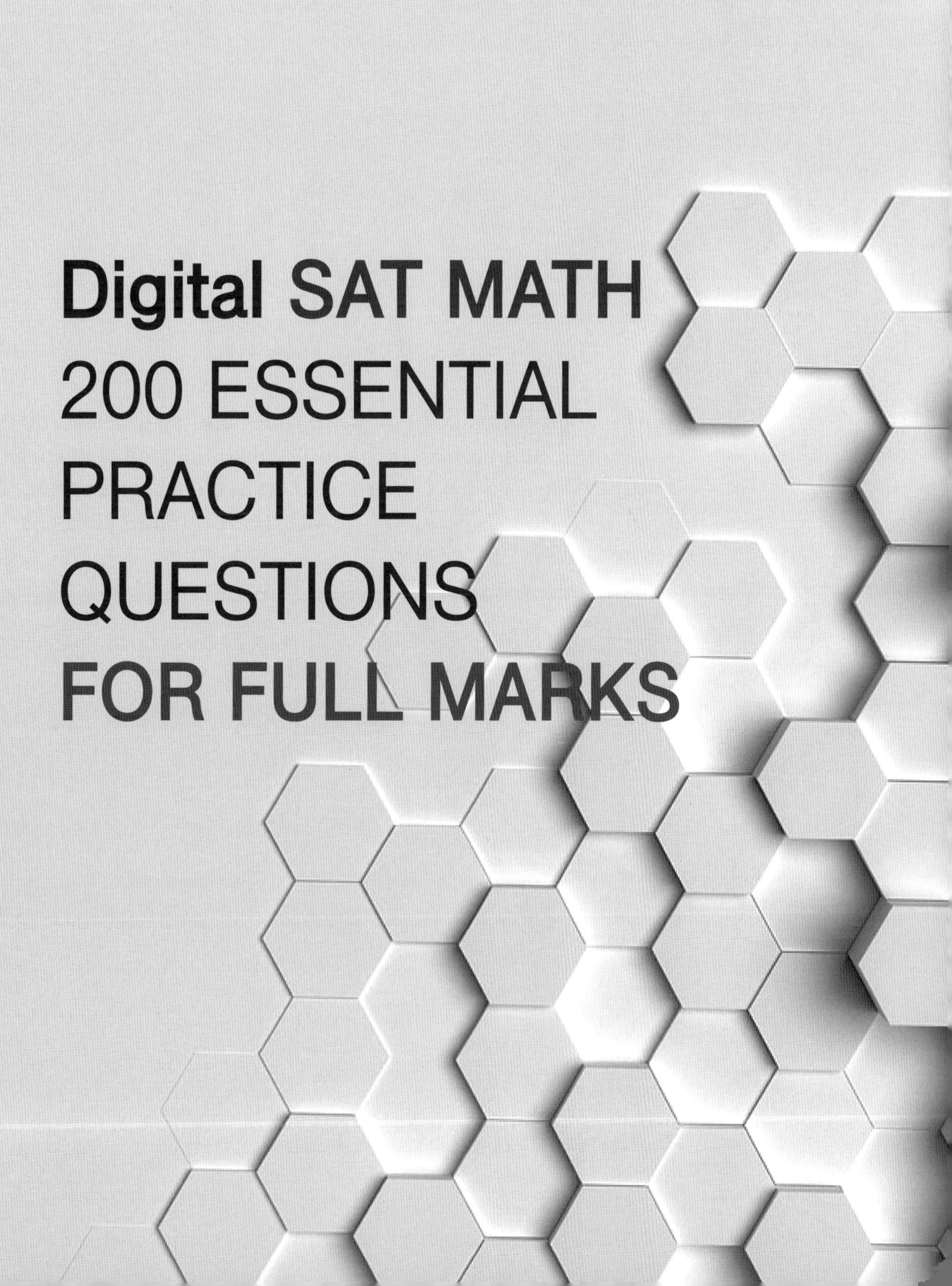

Digital SAT MATH 200 ESSENTIAL PRACTICE QUESTIONS FOR FULL MARKS

Algebra

The chief forms of beauty are older and symmetry and
definiteness, which the mathematical sciences
demonstrate in a specific degree.
-ARISTOTLE (384-322 B.C.)

☑ The Algebra section of the SAT Math

"Algebra" is one of the major content areas tested in the SAT Math section. It focuses on assessing a student's understanding and ability to work with linear equations, linear inequalities, and systems of linear equations. This content area tests fundamental algebraic concepts and serves as the foundation for more complex mathematical concepts.

The Algebra section of the SAT Math primarily evaluates the following key skills:

1. Solving Linear Equations : This skill involves solving equations of the form $ax+b=c$, where a, b, and c are constants and x is the variable. Students are expected to solve for x by performing algebraic operations such as addition, subtraction, multiplication, and division.

2. Interpreting Linear Functions : This skill requires students to understand the properties and behavior of linear functions represented in various forms, such as slope−intercept form ($y=mx+b$) or standard form ($ax+by=c$). They should be able to analyze and interpret the slope, intercepts, and transformations of linear functions.

3. Graphing Linear Equations : Students should be able to graph linear equations and identify key characteristics of the resulting line, such as slope, y−intercept, x−intercept, and slope−intercept form.

4. Solving Systems of Linear Equations : This skill involves solving systems of equations, which are sets of two or more equations with multiple variables. Students are expected to find the values of the variables that satisfy all the equations simultaneously, using methods such as substitution, elimination, or graphing.

5. Solving Linear Inequalities : Students should be able to solve linear inequalities, which involve statements of inequality ($<$, $>$, $\leq$, $\geq$) instead of equality ($=$). They must find the ranges of values that satisfy the given inequalities and represent them on a number line or graphically.

Questions in the Algebra section on SAT Math can range from straightforward equation−solving problems to more complex word problems that require applying algebraic concepts to real−world scenarios.

Preparing for the Algebra section involves practicing a wide range of algebraic problems, understanding the properties and behaviors of linear functions, and developing problem−solving strategies for linear equations and inequalities. By mastering these fundamental algebraic skills, you'll be better equipped to tackle the algebraic concepts tested on the SAT Math section.

1.1 Solving Linear Equations

On the SAT Math section, you may encounter questions that require you to solve linear equations. Linear equations involve variables raised to the first power and have a standard form of

$$ax+b=c$$

where a, b and c are constants and x represents the variable.

Here's a step-by-step guide on how to solve linear equations on the SAT Math:

1. Start by simplifying both sides of the equation, if possible. Combine like terms and use the distributive property to eliminate parentheses.

2. If there are any fractions in the equation, eliminate them by multiplying every term by the least common denominator (LCD) to clear the fractions.

3. Next, isolate the variable x on one side of the equation. To do this, perform inverse operations to move the constants to the other side of the equation. Use addition or subtraction to move terms, and use multiplication or division to cancel out coefficients.

4. Keep performing inverse operations until you have isolated the variable on one side and simplified the other side of the equation.

5. If necessary, simplify further by combining like terms.

6. Once you have isolated the variable, you can determine its value. If the equation is in the form $ax+b=c$ where a, b and c are integers, you can solve for "x" by subtracting "b" from both sides and then dividing both sides by "a".

7. Check your solution by substituting the value of "x" back into the original equation. Ensure that both sides of the equation are equal when the value of "x" is plugged in.

8. It's important to be cautious while solving equations, especially when performing operations like division or multiplication. Be sure to apply the operation to both sides of the equation to maintain equality.

9. Practicing various linear equation problems will help you become more comfortable and efficient at solving them during the SAT Math section.

On the SAT Math section, you may also encounter questions related to systems of linear equations. A system of linear equations consists of multiple equations with the same variables. The main task is to find the values of the variables that satisfy all the given equations simultaneously. Systems of linear equations can be classified based on their solutions. Let's discuss the classifications:

1. Consistent Independent : In this case, the system has a unique solution. It means there is a specific set of values for the variables that satisfy all the equations in the system. Geometrically, this represents the intersection point of the lines representing the equations on a coordinate plane.

2. Consistent Dependent : In this case, the system has infinitely many solutions. It means that any combination of values that satisfies one equation will satisfy all the equations in the system. Geometrically, this represents overlapping lines or coinciding lines on a coordinate plane.

3. Inconsistent : In this case, the system has no solution. It means that there are no values for the variables that satisfy all the equations in the system. Geometrically, this represents parallel lines on a coordinate plane that do not intersect.

When solving a system of linear equations, there are multiple methods you can use, such as substitution, elimination, or matrix methods. The SAT Math section typically tests your ability to choose an appropriate method and efficiently solve the system to determine its classification.

Practice solving a variety of system of linear equations problems to become familiar with the different types of solutions and develop your problem-solving skills for the SAT Math section.

1. LEVEL:3

$$4+5m=4m+1+m$$

Which of the following best describes the solution set to the equation shown above?

Ⓐ The equation has exactly one solution, $m=0$.

Ⓑ The equation has exactly one solution, $m=1$.

Ⓒ The equation has no solutions.

Ⓓ The equation has infinitely many solutions.

2. LEVEL:4

$$-4+bx=2x+3(x+1)$$

In the equation shown above, b is a constant. For what value of b does the equation have no solutions?

Ⓐ 3

Ⓑ 4

Ⓒ 5

Ⓓ 6

1.2 Interpreting Linear Functions : Slope and Intercepts

1. Slope of Line

The slope of the line joining $A(x_1, y_1)$ to $B(x_2, y_2)$

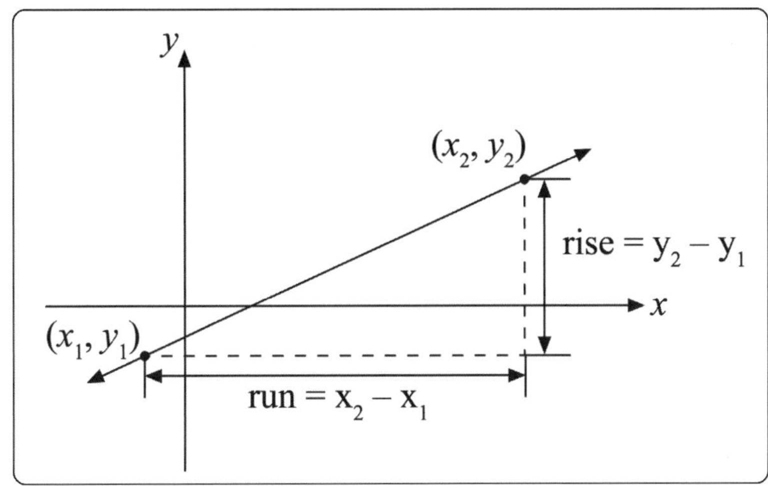

$$slope = \frac{y_2 - y_1}{x_2 - x_1}$$

2. The equation $y = mx + b$

① slope : m ② y−Intercept : b ③ x−Intercept : $-\frac{b}{m}$

On the SAT Math section, you may encounter questions that require you to interpret linear functions. Interpreting linear functions involves understanding and analyzing various components of a linear equation, such as slope, y−intercept, and the relationship between variables. Here's an explanation of the key parts to interpret in linear functions on the SAT Math:

$$y = mx + b$$

1) Slope (m): The slope of a linear function represents the rate of change or the steepness of the line. It determines how much the y−value (vertical axis) changes for a given change in the x−value (horizontal axis). The slope is typically denoted by the letter "m" in the equation $y = mx + b$. A positive slope indicates an upward slope, while a negative slope indicates a downward slope. A slope of zero represents a horizontal line.

2) y−intercept (b) : The y−intercept is the point where the line crosses or intersects the y−axis. It is the value of y when x equals zero. In the equation y = mx + b, the y−intercept is represented by the constant term "b." The y−intercept provides information about the initial value or starting point of the function.

3) Rate of Change : The rate of change, often determined by the slope, describes how much one variable changes with respect to a change in the other variable. For linear functions, the rate of change is constant throughout the entire line. It can be interpreted as the "rise over run" or the change in y divided by the change in x.

4) Linearity: Linear functions have a constant rate of change and produce a straight line when graphed on a coordinate plane. The relationship between the variables remains consistent and follows a linear pattern. This linearity can be used to make predictions or analyze the behavior of the function.

When encountering questions related to interpreting linear functions on the SAT Math section, you might be asked to analyze the slope, determine the y-intercept, identify the rate of change, predict outcomes based on the function, or compare and contrast different linear functions.

To excel in interpreting linear functions, it's crucial to understand the concepts of slope, y-intercept, and rate of change, and practice analyzing linear equations and their graphs. This will enable you to effectively interpret and solve related problems on the SAT Math section.

3. LEVEL: 1

The equation $y=2x+3$ represents a linear function. What does the coefficient of x in this equation represent?

Ⓐ The initial value of x.
Ⓑ The rate of change in y.
Ⓒ The rate of change in x.
Ⓓ The initial value of y.

4. LEVEL: 1

The equation $y=-5$ represents a linear function. What does the constant term in this equation represent?

Ⓐ The initial value of x.
Ⓑ The rate of change in y.
Ⓒ The rate of change in x.
Ⓓ The initial value of y.

5. LEVEL: 2

The equation $y=-0.5x+4$ represents a linear function. What does the y-intercept in this equation represent?

Ⓐ The initial value of x.
Ⓑ The total value of y.
Ⓒ The rate of change in x.
Ⓓ The value of y when x equals zero.

6. LEVEL: 2

The equation $2x+3y=12$ represents a linear function. What does the coefficient of y in this equation represent in term of right handed side, 12?

Ⓐ The initial value of x.
Ⓑ The rate of change in y.
Ⓒ The rate of change in x.
Ⓓ The initial value of y.

7. LEVEL: 3

The equation $y=4x^2+2x-3$ does not represent a linear function. Why is this equation not linear?

Ⓐ It contains an exponent greater than 1.
Ⓑ It does not have a constant term.
Ⓒ It has two variables.
Ⓓ It is not a straight line.

8. LEVEL: 3

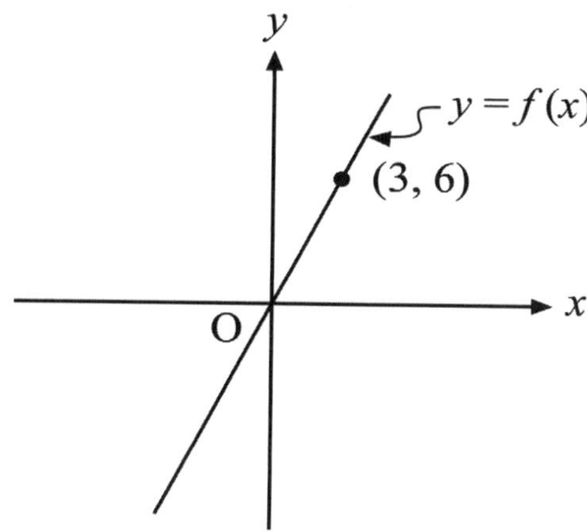

In the xy–plane above, apoint (not shown) with coordinate (s, t) lies on the graph of the linear function f. If s and t are positive integers, what is the ratio of t to s?

Ⓐ 1 to 3
Ⓑ 1 to 2
Ⓒ 2 to 1
Ⓓ 3 to 1

9. LEVEL:4

$$T = 10m + 40$$

Tim decides to cook a steak. The interior temperature, T, of the steak, in degrees Fahrenheit(°F), after cooking for m minutes is given in the equation above. What does the 10 mean in the equation?

Ⓐ The interior temperature is 10°F when Tim starts cooking the steak.
Ⓑ The interior temperature of the steak increases by 10°F for every minute it is cooked.
Ⓒ The interior temperature of the steak decreases by 10°F for every minute it is cooked.
Ⓓ The interior temperature of the steak will increase a total of 10°F while being cooked.

10.

LEVEL:4

Officials project that between 2010 and 2050, the Sub−Saharan African population will drastically change. The model below gives the projection of the population, P, in thousands, with respect to time, t (provided that 2010 is when $t=0$).

$$P = 175 + \frac{11}{2}t$$

What does the 175 mean in the equation?

Ⓐ In 2010, the population of Sub−Saharan African was 175 thousand.
Ⓑ In 2050, the population of Sub−Saharan African will be 175 thousand.
Ⓒ Between 2010 and 2050, the population of Sub−Saharan African will increase by 175 thousand.
Ⓓ Between 2010 and 2050, the population of Sub−Saharan African will decrease by 175 thousand.

11.

LEVEL:3

$$g(t) = \frac{5(7t-12c)}{2} - 25$$

The number of people who go to a public swimming pool can be modeled by the function g above, where c is a constant and t is the air temperature in degrees Fahrenheit ($^\circ F$) for $70 < t < 100$. If 350 people are predicted to go to the pool when the temperature is $90\,^\circ F$, what is the value of c?

Ⓐ 20
Ⓑ 40
Ⓒ 60
Ⓓ 80

12. LEVEL:5

$$35A + 28P = 250$$

Vera is hiring entertainers for her charity event. She is using the equation above to determine the number of hours, A and P, to hire a caricature artist and a face painter for, respectively, with a total budgeted cost of $\$250$. The caricature artist charges $\$35$ per hour and the face painter charges $ 28 per hour. What is the meaning of the $35A$ in this equation?

Ⓐ The caricature artist charges $35A$ dollars per hour.

Ⓑ The caricature artist works for a minimum of $35A$ hours in a week.

Ⓒ For the first 35 hours each week, the caricature artist makes A dollars.

Ⓓ The caricature artist charges a total of $35A$ dollars for A hours worked.

1.3 Linear equation word problem

1. Word Problems Translation

Word Problems Translation Table	
English	**Math**
equals, equivalent to, was, will be, has costs, adds up to, the same as, as much as	$=$
times, of, multiplied by, product of, twice, double, by	$\times$
divided by, per, out of, each, ratio	$\div$
plus, added to, and, sum, combined, total, inceased by	$+$
minus, subtracted from, smaller than, less than, fewer, decreased by, difference between	$-$
a number, how much, how many, what	variable

2. Definition of Absolute Value

Distance from the origin on the number line.

We write $|x|$ to mean the **magnitude** or **modulus** of x.

3. Notice that

(1) $|a| = \begin{cases} a & (a \geq 0) \\ -a & (a < 0) \end{cases}$

(2) $|a| \geqq 0$, $|a| = |-a|$, $|a|^2 = a^2$

(3) $|a||b| = |ab|$, $\dfrac{|a|}{|b|} = \left|\dfrac{a}{b}\right|$

(4) Inequalities

- $|x| < a \Leftrightarrow -a < x < a$
- $|x| > a \Leftrightarrow x > a$ or $x < -a$

13. LEVEL:1

The sum of two numbers is 12, and their difference is 4. Find the two numbers.

14. LEVEL:1

A school is organizing a field trip for 60 students. There are two types of tickets available : adult tickets cost $\$15$ each, and student tickets cost $\$8$ each. If the total cost of the tickets is $\$750$, how many adult tickets were sold? Round to the nearest whole number.

15. LEVEL:2

A store sells two types of laptops: brand A and brand B. The store sold a total of 80 laptops, and the revenue from the sales was $60,000. Brand A laptops sell for $800 each, and brand B laptops sell for $1,000 each. If the store sold twice as many brand A laptops as brand B laptops, how many brand B laptops were sold? Round to the nearest whole number.

16. LEVEL:2

A factory produces two types of products: Product A and Product B. The factory produces a total of 500 products each day. Product A requires 2 hours of labor to produce, while Product B requires 3 hours. Each day, a total of 1,200 hours of labor are available. How many units of Product A and Product B should be produced to maximize production?

17. LEVEL:2

Joanne and Richard volunteer at a hospital. Joanne volunteers 4 hours more per week than Richard does. In a given week, they do not volunteer for more than a combined total of 16 hours. If x is the number of hours that Richard volunteers, which inequality best models this situation?

Ⓐ $x+4 \le 16$

Ⓑ $2x+4 \le 16$

Ⓒ $2x+8 \le 16$

Ⓓ $2x-4 \le 16$

18. LEVEL:2

Maria plans to rent a boat. The boat rental costs $\$60$ per hour, and she will also have to pay for a water safety course that costs $\$10$. Maria wants to spend no more than $\$280$ for the rental anf thr course. If the boat rental is availabe only for a whole number of hours, what is the maximum number of hours for which Maria can rent the boat?

19. LEVEL:3

$$D = 60 - \frac{3}{4}P$$

$$S = \frac{1}{4}P$$

In economics, the equilibrium price is defined as the price at which quantity demanded and quantity supplied are equal. If the quantity demanded, D, and quantity supplied, S, in terms of the price in dollars, P, equilibrium price?

Ⓐ $0

Ⓑ $60

Ⓒ $80

Ⓓ $120

20. LEVEL:3

Mikal has a summer project in which he must complete at least 35 hours of community service at a city park. Each day that he goes to the park, he volunteers for 7 hours. It takes him 1.5 hours to get to the park each way, which also counts toward his community service hours. Which of the following inequalities can be used to find the number of days, d, Mikal must volunteer at the park to complete his summer project?

Ⓐ $7d > 35$

Ⓑ $7d \geq 35$

Ⓒ $7d + 3d \leq 35$

Ⓓ $7d + 3d \geq 35$

21. LEVEL:4

A meteorologist estimates that on a sunny day, the air temperature decreases by about $4\ °F$ for every 1,000 feet (ft) of elevation gain. On a certain day, the air temperature outside an airplane flying above Seattle is $-58\ °F$, and the ground level temperature in Seattle is $70\ °F$. If x is the height, in feet, at which the plane is flying, which of the following best models the situation?

Ⓐ $70=-\dfrac{4}{1000}x-58$

Ⓑ $70=\dfrac{4}{1000}x-58$

Ⓒ $-58=-4x+70$

Ⓓ $-58=4x+70$

1.4 Graphing Linear Equations

1. Slope / y−Intercept Form : $y = ax + b$

where a is the slope of the line and b is the y−intercept of the line.
Both are constants.

2. Slope / x−Intercept Form : $y = m(x - a)$

where m is the slope of the line and b is the x−intercept of the line.
Both are constants.

3. Point/Slope Form : $y - y_1 = m(x - x_1)$

where m is the slope of the line and (x_1, y_1) is a point on the line.

4. Point/Point Form

When (x_1, y_1), (x_2, y_2) is a point on the line.

$$x_1 \neq x_2 : y - y_1 = \frac{y_2 - y_1}{x_2 - x_1}(x - x_1)$$

$$x_1 = x_2 : x = x_1 \text{ or } x_1 = x_2$$

5. Parallel and Perpendicular lines

we need to discuss in this section is that of parallel and perpendicular lines.
Here is a sketch of parallel and perpendicular lines.

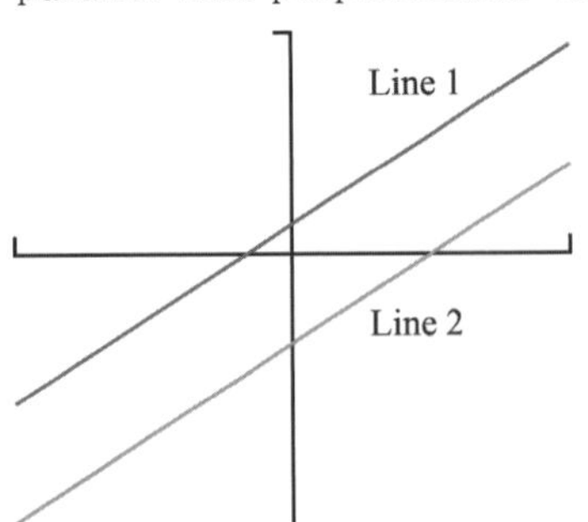

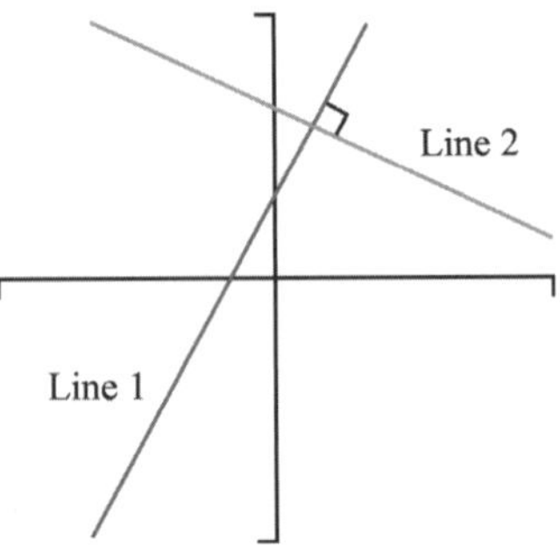

Suppose that the slope of Line 1 is m_1 and the slope of Line 2 is m_2 . We can relate the slopes of parallel lines and we can relate slopes of perpendicular lines as follows.

Parallel : $m_1 = m_2$

Perpendicular : $m_1 m_2 = -1$

22.

LEVEL:2

Which of the following is an equation of the line A graphed in the xy−plane that passes through the point $(-1, 3.5)$ and is perpendicular to the line B whose equation is $x+4.5=0$?

Ⓐ $x = 1$

Ⓑ $x = 3.5$

Ⓒ $y = 3.5$

Ⓓ $y = 4.5$

23.

LEVEL:2

Which of the following equations represents a line in the xy−plane with an x−intercept at $(-2,0)$ and a slope of 4?

Ⓐ $y = 4x + 8$

Ⓑ $y = 4x - 8$

Ⓒ $y = 4x - 2$

Ⓓ $y = 4x + 2$

24.

LEVEL:2

Which of the following equations represents a line in the xy−plane with an x−intercept at $(-15,0)$ and a y−intercept at $(0, -9)$?

Ⓐ $\frac{x}{15} + \frac{y}{9} = 1$

Ⓑ $-\frac{x}{15} - \frac{y}{9} = 1$

Ⓒ $\frac{x}{9} - \frac{y}{15} = 1$

Ⓓ $-\frac{x}{9} + \frac{y}{15} = 1$

25. LEVEL:3

A laser technician sets up a laser at $(a, 2)$ and a target at $(3, a)$ in the xy-plane. If the resulting laser beam is perpendicular to another beam modeled by the line $2x-3y=-8$, What is the value of a?

26. LEVEL:3

The equation for the gravitational potential energy U of a 1-kilogram object on Earth resting h meters above the ground t seconds after placement is $U=9.8h$ joules. Which of the following is a graph of U versus t for a 1-kilogram object 2 meters above the ground?

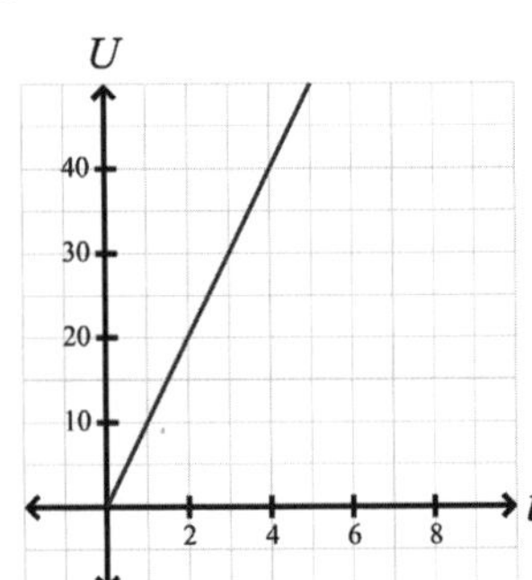

Ⓑ

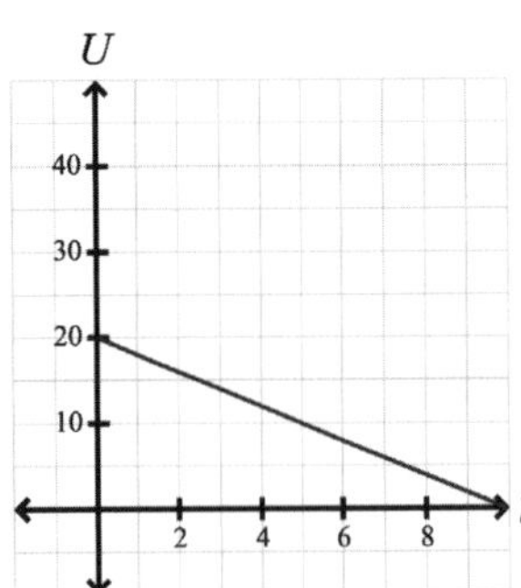

Ⓒ

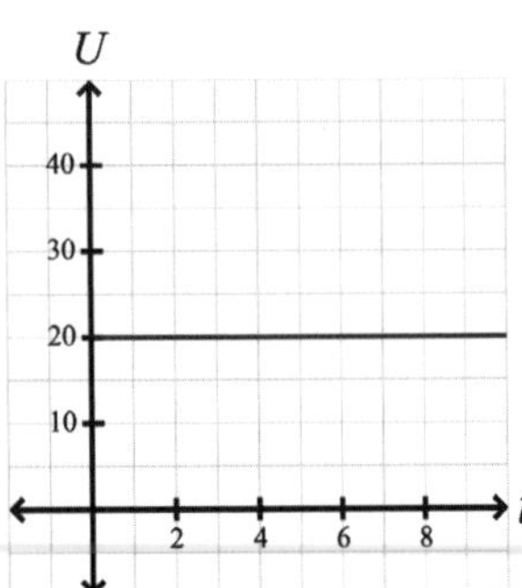

Ⓓ

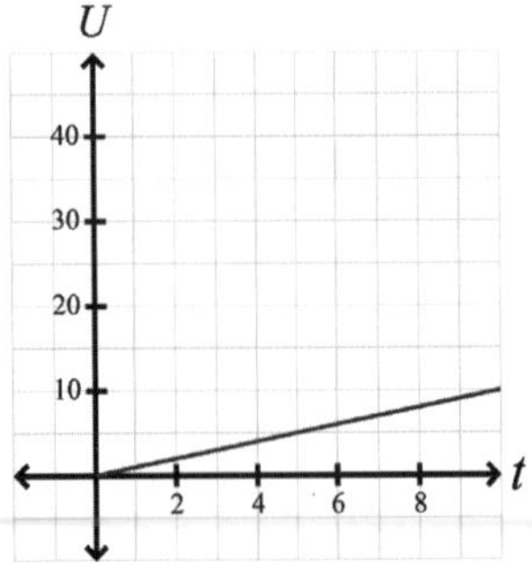

27. LEVEL:3

Two lines graphed in the xy−plane have the equations $2x+5y=20$ and $y=kx-3$, where k is a constant. For what value of k will the two lines be perpendicular?

Ⓐ $-\frac{2}{5}$

Ⓑ $\frac{2}{5}$

Ⓒ $\frac{5}{2}$

Ⓓ $-\frac{5}{2}$

28. LEVEL:4

$$5x + 7y = 1$$
$$ax + by = 1$$

In the given pair of equations, a and b are constant. The graph of this pair of equations in the xy−plane is pair of perpendicular lines. Which of the following pairs of equations also represents a pair of perpendicular lines?

Ⓐ $10x + 7y = 1$
$ax - 2by = 1$

Ⓑ $10x + 7y = 1$
$ax + 2by = 1$

Ⓒ $10x + 7y = 1$
$2ax + by = 1$

Ⓓ $5x - 7y = 1$
$ax + by = 1$

1.5 Linear function word problems

These problems will challenge your ability to apply linear functions to real-world situations and solve for various quantities. Practice setting up and solving linear function word problems to enhance your skills in solving linear function problems on the SAT Math section.

29. LEVEL:1

A car rental company charges a flat fee of \$30 per day plus an additional \$0.25 per mile driven. Write an equation to represent the total cost (C) in dollars of renting a car for a day if x miles are driven.

30. LEVEL:1

The temperature (T) in degrees Celsius increases at a constant rate of 2.5 degrees per hour. If the initial temperature is 18 degrees Celsius, write an equation to represent the temperature (T) after x hours.

31. LEVEL:1

A cell phone plan costs \$20per month plus an additional \$0.10 per minute for calls made. Write an equation to represent the total cost (C) in dollars of the cell phone plan if x minutes are used in a month.

32. LEVEL:2

A roller coaster is currently traveling at a speed of 49 miles per hour (mph). The coaster's speed will increase at a constant rate of 17 mile per hour every 2 hours until the coaster reaches its top speed 5 seconds from now. If $t \leq 5$, which function best represents the roller coaster's speed, in miles per hour, t hours from now?

Ⓐ $f(t) = 49 + 8.5t$
Ⓑ $f(t) = 49 + 17t$
Ⓒ $f(t) = 49t + 9.8t$
Ⓓ $f(t) = 49t + 17t$

33. LEVEL:3

A bamboo plant has a height of 11 feet and grows at a constant rate of 2 feet per day. At this rate, how many days from now will the height of the bamboo plant be 27 feet?

34. LEVEL:3

The boiling point of water at sea level is 211 degrees Fahrenheit($^\circ F$). For every increase of 1000 feet above sea level, the boiling point of water drops approximately $1.84\ ^\circ F$. Which of the following equations gives the approximate boiling point B, in $^\circ F$, at h teet above sea level?

Ⓐ $B = 211 - 1.84h$
Ⓑ $B = 211 - 0.00184h$
Ⓒ $B = 211h$
Ⓓ $B = 1.84(211) - 1000h$

35. LEVEL:3

Sterling silver is an alloy of silver that is 92.5% pure silver. If x grams of sterling silver are mixed with y grams of an 88% silver alloy to produce a 91% silver alloy, which of the following equations correctly relates x and y?

Ⓐ $0.075x + 0.12y = 0$
Ⓑ $0.015x - 0.03y = 0$
Ⓒ $0.925x + 0.88y = 91$
Ⓓ $0.925x + 0.88y = 0.91xy$

36. LEVEL:4

Gustav is starting a special sprinting regimen in outdoor track. The regimen suggests sprinting for a certain number of seconds, s, at first. The second time he sprints, he should run 150 more seconds than the first time. For the third time, the regimen suggests sprinting twice as the first time. And finally, the regimen suggests an 80 seconds sprint. Which of the following functions can be used to find the total number of seconds, t, suggested by the entire sprinting regimen?

Ⓐ $t = s + 230$
Ⓑ $t = 2s + 230$
Ⓒ $t = \frac{5}{2}s + 230$
Ⓓ $t = 4s + 230$

1.6 Systems of linear inequalities word problems

Systems of linear inequalities word problems on the SAT Math section require you to analyze multiple inequalities and identify the regions on a graph that satisfy all the given conditions. These problems often involve constraints or limitations and ask you to find the feasible region or range of values that satisfy all the conditions simultaneously.

To solve systems of linear inequalities word problems, you typically follow these steps:

1. Identify the variables : Determine the quantities or variables involved in the problem and assign variables to represent them. For example, if the problem involves quantities like the number of items sold or the cost of tickets, assign variables to represent these quantities.

2. Write the inequalities : Translate the given conditions or constraints into inequalities. Each inequality represents a condition or constraint on the variables. Use symbols like "$<$" (less than), "$>$" (greater than), "$\leq$" (less than or equal to), "$\geq$" (greater than or equal to), or "$=$" (equal to).

3. Graph the inequalities : Plot the graphs of each inequality on a coordinate plane. To do this, first rewrite each inequality in slope-intercept form ($y=mx+b$) or standard form ($ax+by=c$) to identify the slope and y-intercept or x-intercept. Then, plot the lines or boundaries represented by the inequalities. Use dashed or solid lines based on the inequality symbols.

4. Determine the feasible region : Identify the overlapping or shaded region on the graph where all the inequalities are satisfied simultaneously. This region represents the feasible solutions or values that satisfy all the given conditions.

5. Interpret the solution : Read the feasible region and interpret it in the context of the problem. Depending on the problem, the feasible region might represent a range of values, restrictions, or limitations on the variables.

6. Answer the question : Based on the feasible region, answer the specific question asked in the problem. It could be finding the maximum or minimum value, determining a specific range, or identifying the number of solutions within the feasible region.

It's important to pay attention to the shading and overlapping areas on the graph to determine the feasible solutions accurately. Also, be mindful of any additional information or restrictions mentioned in the problem that might affect the feasible region.

37. LEVEL:1

A company produces two products: Product A and Product B. Product A requires 4 hours of labor to produce, and Product B requires 3 hours. The company has a total of 60 hours of labor available. Write a system of inequalities to represent the number of units of each product the company can produce.

38. LEVEL:2

A juice shop sells two types of fruit juices: apple juice and orange juice. The shop has 100 liters of apple juice and 150 liters of orange juice available. The apple juice sells for $3 per liter, and the orange juice sells for $2 per liter. The shop wants to earn at least $300 in sales. Write a system of inequalities to represent the number of liters of each juice the shop needs to sell to meet the sales goal.

39. LEVEL:2

A farmer wants to plant two crops: corn and wheat. The farmer has 40 acres of land available. Planting an acre of corn requires 2 hours of labor, and planting an acre of wheat requires 3 hours. The farmer has a total of 100 hours of labor available. Write a system of inequalities to represent the number of acres of each crop the farmer can plant.

40. LEVEL:2

A store sells two types of televisions: Standard and High-definition (HD). The store has a budget of at least $\$3000$ to purchase televisions. Standard televisions cost $\$500$ each, and HD televisions cost $\$800$ each. The store wants to buy at least 4 televisions. Write a system of inequalities to represent the number of each type of television the store can purchase.

41. LEVEL:2

$$y > 2x - 1$$
$$2x > 5$$

Which of the following consists of the y-coordinates of all the points that satisfy the system of inequalities above?

Ⓐ $y > 6$

Ⓑ $y > 4$

Ⓒ $y > \frac{5}{2}$

Ⓓ $y > \frac{3}{2}$

42. LEVEL: 3

Ken is working this summer as part of a crew on a farm. He earned $\$8$ per hour for the first 10 hours he worked this week. Because of his performance, his crew leader raised his salary to $\$10$ per hour for the rest of the week. Ken saves 90% of his earnings from each week. What is the least number of hours he must the rest of the week to save at least $\$270$ for the week?

43. LEVEL: 3

Jackie has two summer jobs. She works as a tutor, which pays $\$12$ per hour, and she works as a lifeguard, which pays $\$9.50$ per hour. She can work no more 20 hours per week, but she wants to earn at least $\$220$ per week. Which of the following systems of inequalities represents this situation in terms of x and y, where x is the number of hours she tutors and y is the number of hours she works as a lifeguard?

Ⓐ $12x + 9.5y \leq 220$
$x + y \geq 20$

Ⓑ $12x + 9.5y \leq 220$
$x + y \leq 20$

Ⓒ $12x + 9.5y \geq 220$
$x + y \leq 20$

Ⓓ $12x + 9.5y \geq 220$
$x + y \geq 20$

44. LEVEL: 4

Students in a science lab are working in groups to build both a small and a large electrical circuit. A large circuit uses 4 resistors and 2 capacitors, and a small circuit uses 3 resistors and 1 capacitor. There are 100 resistors and 70 capacitors available, and each group must have enough resistors and capacitors to make one large and one small circuit. What is the maximum number og groups that could work on this lab project?

45. LEVEL: 4

A business analyst is deciding the amount of time allotted to each employee for meetings and training. He wants the sum of meeting and training time to be no more than 16 hours per month. Also, there should be at least one training hour for every two meeting hours. Finally, there should be at least 2 hours for meeting per month to discuss short−term goals. What is the difference between the maximum and minimum number of monthly training hours that could be allotted to an employee?

Ⓐ 10 *hours*

Ⓑ 13 *hours*

Ⓒ 14 *hours*

Ⓓ 16 *hours*

1.7 Solving systems of linear equations

Solving systems of linear equations on the SAT Math section involves finding the values of variables that satisfy multiple equations simultaneously. These problems may present the equations in various forms, such as general form ($ax+by=c$), slope-intercept form ($y=mx+b$), or a combination of both.

To solve systems of linear equations, you typically follow these steps:
1. Identify the variables: Determine the variables present in the equations. Assign variables, such as x and y, to represent the unknown quantities.

2. Choose a method: There are several methods for solving systems of linear equations, including substitution, elimination, and graphing. Depending on the given equations, choose the most efficient method for solving the system.

- Substitution method: Solve one equation for one variable in terms of the other variable and substitute this expression into the other equation. Solve the resulting equation to find the value of one variable, then substitute it back to find the value of the other variable.

- Elimination method: Multiply one or both equations by appropriate constants to create coefficients that will cancel out when the equations are added or subtracted. Add or subtract the equations to eliminate one variable, and solve the resulting equation to find the value of the remaining variable.

- Graphing method: Graph each equation on a coordinate plane and find the point of intersection, which represents the solution to the system of equations. This method is best suited when the equations are given in slope-intercept form.

3. Solve the system: Apply the chosen method to solve the system of equations. Perform the necessary steps to find the values of the variables that satisfy all the equations simultaneously.

4. Check the solution: Substitute the found values back into the original equations to verify that they satisfy each equation.

5. Interpret the solution: State the values of the variables that represent the solution to the system of equations. Depending on the problem, the solution might be a single point (x, y), an expression in terms of one variable, or a range of values.

6. Answer the question: Based on the solution, answer the specific question asked in the problem. It could involve finding a specific value, determining the number of solutions, or identifying any restrictions or conditions.

It's important to understand the different methods and choose the most appropriate one for each problem. Practice solving systems of linear equations using various methods to develop your skills in solving these types of problems effectively on the SAT Math section.

46. LEVEL:3

$$-5.1x + 3y = 1.2$$
$$3.2x - 8y. = b$$

Which of the following choices of b will result in a system of linear equations with exactly one solution?

Ⓐ b can be any number

Ⓑ b can be any number except 1.2

Ⓒ b can be any number except -1.2

Ⓓ $b = 1.2$

47. LEVEL:4

$$0.6 = 1.5\,(a + c\,(b + 0.8\,))$$
$$-0.2 = -2.5(b - 0.4(1.2 - 1.5a))$$

Consider the system of equations above, where c is a constant. For which value of c are there no (a,b) solutions?

Ⓐ 0

Ⓑ $\frac{5}{17}$

Ⓒ $\frac{5}{3}$

Ⓓ None of the above

Passport to Advanced Math

“

The control of large number is possible and like unto
that of small numbers, if we subdivide them.
-SUN TIZE (544-496 B.C.)

☑ Passport to Advanced Math section of the SAT Math

"Passport to Advanced Math" is one of the content areas tested in the SAT Math section. It assesses a student's ability to work with more complex mathematical concepts, including algebraic expressions, equations, functions, and advanced mathematical operations. This content area focuses on higher-level math skills and serves as a bridge between intermediate algebra and more advanced math topics.

The Passport to Advanced Math section of the SAT Math primarily evaluates the following key skills:

1. Quadratic Equations and Expressions : This skill involves working with quadratic equations and expressions, including factoring, solving quadratic equations using various methods (factoring, completing the square, quadratic formulas, and understanding the properties of quadratic functions.

2. Exponents and Radicals : Students are expected to understand and work with exponential and radical expressions, including simplifying expressions with exponents and radicals, solving equations with exponents and radicals, and applying properties of exponents and radicals.

3. Rational and Radical Equations : This skill involves solving and manipulating rational expressions and equations, including simplifying, multiplying, dividing, adding, and subtracting rational expressions, solving equations with rational expressions, and solving equations with radical expressions.

4. Functions and Function Notation : Students should be familiar with the concept of functions, including identifying and analyzing functions, evaluating functions, and understanding function notation ($f(x)$).

5. Advanced Operations with Polynomials: This skill requires students to work with polynomial expressions, including multiplying, dividing, adding, and subtracting polynomials, and applying polynomial operations to solve equations and evaluate expressions.

6. Complex Numbers: Students should be able to work with complex numbers, including simplifying complex expressions, performing operations with complex numbers (addition, subtraction, multiplication, division), and solving equations involving complex numbers.

Questions in the Passport to Advanced Math section can range from algebraic manipulation and equation-solving problems to more abstract and conceptual questions related to advanced mathematical topics.

Preparing for the Passport to Advanced Math section involves practicing a wide range of advanced algebraic concepts, mastering the properties and behaviors of quadratic functions and exponential functions, and developing problem-solving strategies for complex equations and expressions. By strengthening your skills in these advanced mathematical concepts, you'll be better equipped to tackle the more challenging questions in the SAT Math section.

2.1 Radicals and Rational Exponentials

1. The key components of Radicals and Rational Exponents

In the "Passport to Advanced Math" section of the SAT Math, one of the content areas tested is "Radicals and Rational Exponents." This topic focuses on the understanding and manipulation of expressions involving radicals (square roots, cube roots, etc.) and rational exponents (exponents expressed as fractions). These concepts are essential in working with more advanced algebraic expressions and equations.

1. Simplifying Radicals : Students should be able to simplify radical expressions by identifying perfect square factors and applying the rules of multiplication and division with radicals. This includes simplifying square roots, cube roots, and higher-order roots.

2. Operations with Radicals : Students should be able to perform operations with radical expressions, such as adding, subtracting, multiplying, and dividing radicals. This involves applying the rules of combining like terms and using the properties of radicals.

3. Rational Exponents : Students should understand that rational exponents represent a fractional power. They should be able to rewrite expressions with rational exponents as radical expressions and vice versa. This includes understanding the relationship between fractional exponents and radical notation.

4. Laws of Exponents with Radicals : Students should be familiar with the laws of exponents and how they apply to expressions involving radicals. This includes laws such as the product rule, quotient rule, power rule, and simplifying expressions with negative exponents.

5. Equations and Inequalities with Radicals: Students should be able to solve equations and inequalities involving radicals. This involves isolating the radical expression, squaring both sides (if necessary), and checking for extraneous solutions.

6. Rationalizing Denominators : Students should understand how to rationalize the denominators of expressions that contain radicals. This involves multiplying the numerator and denominator by a suitable expression to eliminate the radical from the denominator.

Questions in the Radicals and Rational Exponents section may require simplifying radical expressions, solving equations involving radicals, converting between radical and exponent notation, or applying the laws of exponents to expressions with radicals.

2. Radicals and The Operations Between Radicals

Radical Expression

$$\sqrt[n]{x^m}$$

n: Index number m : Exponent x : Radicand

1. Basic Definitions of Radical Exponent

(1) $\sqrt[n]{a}=a^{\frac{1}{n}}$ (2) $\sqrt{a}=\sqrt[2]{a}=a^{\frac{1}{2}}$

2. Converting General Radical Form to Exponential Form:

$$\sqrt[n]{a^m}=a^{\frac{m}{n}}$$

where n and m are real numbers.

All square root values can change exponents.

3. Associative : $(\sqrt[n]{a})^m=\sqrt[n]{a^m}=a^{\frac{m}{n}}$

4. Simple Product : $\sqrt[n]{a}\sqrt[n]{b}=\sqrt[n]{ab}$

5. Complex Product : $\sqrt[n]{a}\sqrt[m]{b}=\sqrt[nm]{a^m b^n}$

6. Simple Quatient : $\dfrac{\sqrt[n]{a}}{\sqrt[n]{b}}=\sqrt[n]{\dfrac{a}{b}}$

7. Nesting : $\sqrt[n]{\sqrt[m]{a}}=\sqrt[nm]{a}$

8. Simple Rationalizing denominator :

$$\frac{\sqrt{a}}{\sqrt{b}}=\frac{\sqrt{a}\sqrt{b}}{\sqrt{b}\sqrt{b}}=\frac{\sqrt{ab}}{b}$$

9. Conjugates : Any pair of expressions fitting the form of $a+b$ and $a-b$

10. General Rationalizing denominator :

$$\frac{c}{\sqrt{a}+\sqrt{b}}=\frac{c(\sqrt{a}-\sqrt{b})}{(\sqrt{a}+\sqrt{b})(\sqrt{a}-\sqrt{b})}$$
$$=\frac{c(\sqrt{a}-\sqrt{b})}{a-b}$$

48. LEVEL:1

Simplify the expression : $\sqrt{9}$

49. SLEVEL:1

Evaluate the expression : $(\sqrt[3]{8})^2$

50. LEVEL:1

Simplify the expression : $\sqrt{\frac{16}{25}}$

51. LEVEL:1

Simplify the expression: $(\sqrt{3} - \sqrt{2})(\sqrt{3} + \sqrt{2})$

52. LEVEL:2

Solve the equation: $\sqrt[3]{x-1} = 2$

53. LEVEL:2

$$\frac{\sqrt{2}}{2}(\sqrt{8} - \sqrt{50})$$

What is the value of the above expression?

Ⓐ -3

Ⓑ -6

Ⓒ $-2\sqrt{21}$

Ⓓ $4 - \sqrt{5}$

54. LEVEL:2

$$\frac{8^{\frac{1}{2}}}{2^{\frac{1}{3}}}$$

Which of the following expressions is equivalent to the expression above?

Ⓐ $2^{\frac{7}{6}}$

Ⓑ $2^{\frac{9}{2}}$

Ⓒ $4^{\frac{1}{6}}$

Ⓓ $4^{\frac{3}{2}}$

55. LEVEL:2

$$(2b^{-5})^3$$

Which of the following is equivalent to the above expression for $b \neq 0$?

Ⓐ $\dfrac{2}{b^{15}}$

Ⓑ $\dfrac{8}{b^{15}}$

Ⓒ $\dfrac{1}{2b^{15}}$

Ⓓ $\dfrac{1}{8b^{15}}$

56. LEVEL:3

Which of the following expressions is equivalent to $(16x^2)^{\frac{1}{2}}$?

Ⓐ $4|x|$

Ⓑ $8|x|$

Ⓒ $\sqrt{8x}$

Ⓓ $16x$

2.2 Operations with irrational expressions

"Operations with Irrational Expressions" is a concept covered in the "Passport to Advanced Math" section of the SAT Math. It involves working with expressions that contain irrational numbers, such as square roots or other non−repeating and non−terminating decimals. This topic focuses on performing operations such as addition, subtraction, multiplication, and division with these irrational expressions.

1. Simplifying Irrational Expressions : Students should be able to simplify expressions that involve irrational numbers. This includes simplifying square roots, cube roots, or other radicals by identifying perfect square factors and applying the rules of multiplication, division, and simplification.

2. Adding and Subtracting Irrational Expressions : Students should be able to add or subtract expressions that contain irrational numbers. This involves combining like terms and simplifying the resulting expression.

3. Multiplying and Dividing Irrational Expressions : Students should be able to multiply or divide expressions involving irrational numbers. This includes applying the distributive property, simplifying radicals, and rationalizing denominators if necessary.

4. Rationalizing Denominators: Students should understand how to rationalize the denominators of expressions that contain irrational numbers. This involves multiplying the numerator and denominator by a suitable expression to eliminate the radical or irrational term from the denominator.

5. Simplifying Complex Expressions: Students may encounter more complex expressions involving both irrational and rational terms. They should be able to simplify and combine these terms by applying the rules of arithmetic and simplification.

Questions in the Operations with Irrational Expressions section may require simplifying or performing operations with expressions containing square roots, cube roots, or other irrational numbers. These questions may involve combining like terms, simplifying radicals, or rationalizing denominators.

57.

LEVEL:3

$$\sqrt{x} \times \sqrt{\frac{y^5}{x^3}}$$

For $x, y \geq 0$, which of the following is equivalent to the above expression?

Ⓐ $x^{-2}y^5$

Ⓑ $x^{-1}y^{\frac{5}{2}}$

Ⓒ $x\,y^{\frac{5}{2}}$

Ⓓ $\dfrac{y^{\sqrt{5}}}{x^{\sqrt{2}}}$

58.

LEVEL:4

The quotient of $\dfrac{\sqrt[5]{m^4}}{\sqrt[4]{m^3}}$ and $\dfrac{\sqrt{m}}{\sqrt[5]{m^6}}$ equals m^y for some real y. What is the value of y?

59. LEVEL:4

$$\frac{\sqrt{x}}{5\sqrt[5]{x^6}}$$

Which of the following expressions is equivalent to the expression above?

Ⓐ $\dfrac{1}{5\sqrt[6]{x}}$

Ⓑ $\dfrac{1}{5\sqrt[10]{x^7}}$

Ⓒ $\dfrac{1}{5\sqrt[10]{x^3}}$

Ⓓ $\dfrac{\sqrt[7]{x}}{5}$

60. LEVEL:4

$$\sqrt{2x^2+7}+5=0$$

How many real solutions does the above equation have?

Ⓐ 0

Ⓑ 1

Ⓒ 2

Ⓓ more than 2

2.3 Rational Operation and Equations

1. The key components of Rational Operations and Equations

"Rational Operations and Equations" is a concept covered in the "Passport to Advanced Math" section of the SAT Math. It involves working with rational expressions, which are expressions that involve fractions with variables in the numerator and/or denominator. This topic focuses on performing operations such as addition, subtraction, multiplication, and division with these rational expressions, as well as solving equations that involve rational expressions.

Here are the key components of Rational Operations and Equations in the context of the SAT Math:

1) Simplifying Rational Expressions : Students should be able to simplify rational expressions by factoring both the numerator and denominator and canceling out common factors. This includes simplifying complex fractions and identifying restrictions on the variables to avoid division by zero.

2) Adding and Subtracting Rational Expressions : Students should be able to add or subtract rational expressions. This involves finding a common denominator, combining like terms, and simplifying the resulting expression.

3) Multiplying and Dividing Rational Expressions : Students should be able to multiply or divide rational expressions. This includes multiplying numerators and denominators, simplifying the resulting expression, and canceling out common factors.

4) Solving Rational Equations : Students should be able to solve equations that involve rational expressions. This includes finding the common denominator, cross-multiplying, and solving for the variable. It's important to check for extraneous solutions that may arise from simplifying and canceling out terms.

5) Complex Fraction Manipulation : Students may encounter complex fractions within rational expressions. They should be able to simplify and manipulate these complex fractions by using techniques such as multiplying by the reciprocal or finding a common denominator.

2. Rational Expressions And Rule for Fractions

1) Fraction Equality :

$$\frac{a}{b}=\frac{c}{d} \Leftrightarrow a : b=c : d \Leftrightarrow ad=bc$$

2) Proportional Rules :

If $a=c$,

then $\frac{a+b}{b}=\frac{c+d}{d}$, $\frac{a-b}{b}=\frac{c-d}{d}$ and $\frac{a+b}{a-b}=\frac{c+d}{c-d}$, $\frac{a-b}{a+b}=\frac{c-d}{c+d}$ (All Denominator $\neq 0$)

3) Componendo

If $\frac{a}{b}=\frac{c}{d}=\frac{e}{f}$, then $\frac{a}{b}=\frac{c}{d}=\frac{e}{f}=\frac{a+c+e}{b+d+f}=\frac{pa+qc+re}{pb+qd+rf}$

($b+d+f \neq 0$, $pb+qd+rf \neq 0$)

4) Dividing rational expressions.

$$\frac{\frac{a}{b}}{\frac{c}{d}}=\frac{a}{b} \div \frac{c}{d}=\frac{a}{b} \times \frac{d}{c}=\frac{ad}{bc}$$

61.

LEVEL:4

$$\frac{2x+6}{(x+2)^2}-\frac{2}{x+2}$$

The expression above is equivalent to $\frac{a}{(x+2)^2}$ where a is a positive constant and $x \neq -2$.

What is the value of a ?

62.

LEVEL:4

$$\frac{\frac{(64b^2-25)}{2b+b^2}}{\frac{25-40b}{b+2}}$$

Which expression is equivalent to the above quotient for all $b > 2$?

Ⓐ $-\frac{8b+5}{5b}$

Ⓑ $\frac{8+5b}{5b}$

Ⓒ $-\frac{5(8-5b)^3}{b(b+2)^2}$

Ⓓ -8

63. LEVEL:4

$$\frac{\dfrac{(3x^2-2x-5)}{x^2+x}}{\dfrac{15x-9}{3x}}$$

Which expression is equivalent to the above quotient for all $x < -2$?

Ⓐ $-\dfrac{x-1}{x+1}$

Ⓑ $\dfrac{3x-5}{5x-3}$

Ⓒ $\dfrac{5x-3}{3x-5}$

Ⓓ $\dfrac{(3x-5)(5x-3)}{x^2}$

2.4 Quadratic factoring and Solving quadratic equations

1. Expansion vs Quadratic factoring

Before proceeding with this section we should note that the topic of solving quadratic equations will be covered in two sections. This is done for the benefit of those viewing the material on the web. This is a long topic and to keep page load times down to a minimum the material was split into two sections.

So, we are now going to solve quadratic equations. First, the standard form of a quadratic equation is

$$ax^2+bx+c=0$$

The only requirement here is that we have an x^2 n the equation. We guarantee that this term will be present in the equation by requiring $a \neq 0$. Note however, that it is okay if b and/or c are zero.

2. Solving by Factoring

As the heading suggests we will be solving quadratic equations here by factoring them. To do this we will need the following fact.

$$\text{If } ab=0 \text{ then either } a=0 \text{ and/or } b=0$$

Before actually starting this discussion we need to recall the distributive law. This will be used repeatedly in the remainder of this section. Here is the distributive law.

$$a(b+c)=ab+ac$$

And another formulas of this example all use one of the following special products.

$$(a+b)(a-b)=a^2-b^2$$
$$(a+b)^2=a^2+2ab+b^2$$
$$(a-b)^2=a^2-2ab+b^2$$

Be careful to not make the following mistakes!

$$(a+b)^2 \neq a^2+b^2$$
$$(a-b)^2 \neq a^2-b^2$$

Also important another formula of the example all use one of the following special products.

$$(ax+b)(cx+d)=acx^2+adx+bcx+bd$$

64. LEVEL:2

Factor by grouping each of the following.

(1) $3x^2-2x+12x-8$

(2) x^5+x-2x^4-2

(3) $x^5-3x^3-2x^2+6$

65. LEVEL:2

Factor each of the following polynomials.

(1) $x^2+2x-15$

(2) $x^2-10x+24$

(3) x^2+6x+9

(4) $3x^2+2x-8$

(5) $5x^2-17x+6$

(6) $4x^2+10x-6$

2.5 Quadratic Equations And Quadratic Formula

This is the final method for solving quadratic equations and will always work. Not only that, but if you can remember the formula it's a fairly simple process as well.
We can derive the quadratic formula by completing the square on the general quadratic formula in standard form. Let's do that and we'll take it kind of slow to make sure all the steps are clear.
First, we MUST have the quadratic equation in standard form as already noted. Next, we need to divide both sides by a to get a coefficient of one on the x^2 term.

$$ax^2+bx+c=0$$

$$x^2+\frac{b}{a}x+\frac{c}{a}=0$$

Next, move the constant to the right side of the equation.

$$x^2+\frac{b}{a}x=-\frac{c}{a}$$

Now, we need to compute the number we'll need to complete the square. Again, this is one-half the coefficient of x squared.

$$\left(\frac{b}{2a}\right)^2=\frac{b^2}{4a^2}$$

Now, add this to both sides, complete the square and get common denominators on the right side to simplify things up a little.

$$x^2+\frac{b}{a}x+\frac{b^2}{4a^2}=-\frac{c}{a}+\frac{b^2}{4a^2}$$

$$\left(x+\frac{b}{2a}\right)^2=\frac{b^2-4ac}{4a^2}$$

Now we can use the square root property on this.

$$x+\frac{b}{2a}=\sqrt{\frac{b^2-4ac}{4a^2}}$$

Solve for x we'll also simplify the square root a little.

$$x=-\frac{b}{2a}+\sqrt{\frac{b^2-4ac}{4a^2}}$$

As a last step we will notice that we've got common denominators on the two terms and so we'll add them up. Doing this gives the following formula.

Quadratic Formula

$$x=-\frac{b+\sqrt{b^2-4ac}}{2a}$$

66. LEVEL:2

$$(v+\frac{1}{5})^2-9=0$$

What is the sum of the solutions to the equation above?

Ⓐ $-\frac{3}{5}$

Ⓑ $-\frac{2}{5}$

Ⓒ $-\frac{1}{5}$

Ⓓ 0

67. LEVEL:2

Use the quadratic formula to solve each of the following equations.

(1) $x^2+2x=7$

(2) $7t^2=6-19t$

(3) $\frac{3}{y-2}=\frac{1}{y}+1$

(4) $16x-x^2=0$

68. LEVEL:4

$$\frac{x}{x-1}+\frac{4}{x-2}=\frac{4}{(x-1)(x-2)}$$

What is the sum of all the possible values for x that satisfy the equation above?

Ⓐ -4

Ⓑ -2

Ⓒ 1

Ⓓ 2

69. LEVEL:4

$$\frac{2x}{x+3} - \frac{4}{x-2} = -3$$

What is the sum for all the values of x that satisfy the equation above?

Sum and Products of Roots on Quardratic Equation

$ax^2+bx+c=0$ has two roots $x=\alpha$ and $x=\beta$ then
$ax^2+bx+c=a(x-\alpha)(x-\beta)$ where $a \neq 0$

1. Sum of Roots : $\alpha+\beta=-\frac{b}{a}$

2. product of Roots : $\alpha\beta=\frac{c}{a}$

70. LEVEL:3

$$4+kp=4p^2$$

In the equation above, k is a constant. If the sum of the solutions to the equation is 0, what is the value of k?

Ⓐ 0

Ⓑ 1

Ⓒ 2

Ⓓ 4

2.6 Quadratic Functions

1. The important properties of Quadratic function for SAT Math:

A quadratic function is a type of polynomial function with a degree of 2. It is expressed in the form

$$f(x) = ax^2 + bx + c$$

where 'a', 'b', and 'c' are constants. The coefficient 'a' cannot be zero, as that would result in a linear function.

- Vertex : The vertex is the point on the graph of a quadratic function where the function reaches its minimum (if 'a' is positive) or maximum (if 'a' is negative) value. The x−coordinate of the vertex can be found using the formula $x = -\frac{b}{2a}$, and the y−coordinate can be obtained by substituting the x−coordinate into the function.

- Axis of Symmetry : The axis of symmetry is a vertical line that passes through the vertex, dividing the graph into two symmetrical halves. The equation of the axis of symmetry is $x = -\frac{b}{2a}$.

- Discriminant : The discriminant is a value that helps determine the nature of the solutions (or roots) of a quadratic equation. It is calculated using the formula $D = b^2 - 4ac$. The discriminant can be used to identify whether the quadratic equation has two real solutions ($D > 0$), one real solution ($D = 0$), or no real solutions ($D < 0$).

- Vertex Form : The vertex form of a quadratic function is given by $f(x) = a(x-h)^2 + k$, where (h, k) represents the vertex of the parabola. This form provides useful information about the vertex and allows for easy graphing and identification of transformations.

- Intercepts : The x−intercepts (or roots) of a quadratic function are the values of 'x' where the graph intersects the x−axis. They can be found by solving the quadratic equation $ax^2 + bx + c = 0$. The y−intercept is the value of 'y' when 'x' is zero and can be obtained by substituting $x = 0$ into the function.

2. Standard form of Quardratic Function

$$y = a(x - h)^2 + k$$

where a, h and k are constants where $a \neq 0$.

(1) The vertex : (h, k)

(2) The axis of symmetry : $x = h$

(3) The parabola opens upward if $a > 0$ and downward if $a < 0$.

3. General Form

The general form of the equation of a parabola is:

$$y = ax^2 + bx + c$$

where a, b and c are constants.

(1) The vertex : $\left(-\frac{b}{2a}, -\frac{b^2 - 4ac}{4a}\right)$

(2) The axis of symmetry : $x = -\frac{b}{2a}$

(3) The parabola opens upward if $a > 0$ and downward if $a < 0$.

(4) The y-intercept is the point $(0, c)$.

71. LEVEL:2

Which of the following equations has a graph in the xy-plane with no x-intercepts?

Ⓐ $y = x^2 + 3x + 4$

Ⓑ $y = x^2 - 5x - 6$

Ⓒ $y = 3x^2$

Ⓓ $y = 2x - 5$

72. LEVEL:3

A minor league hockey team has been collecting ticket sales data over the past year. At a current price of $\$25$ per ticket, an average of 4000 seats are purchased. They predict that for each $\$1$ increase in ticket price, 100 fewer tickets will be sold. Which of the following functions best models the amount of money that the hockey teams expect to collect from ticket sales, y, based on an $\$x$ increase in ticket price?

Ⓐ $y = (25 + x)(4000 - 100x)$

Ⓑ $y = (25 - x)(4000 + 100x)$

Ⓒ $y = x(4000 - 100x)$

Ⓓ $y = 4000(25 + x)$

73. LEVEL:3

Dalia tosses a ball to Kaylee. The ball travels along the path of a parabola and reaches a maximum height of $10ft$ above ground level after traveling a horizontal distance of $5ft$. Let x represent the horizontal distance the ball has traveled, and let y represent the height of the ball above ground level. If Dalia releases the ball at an initial height of $3ft$, which of the following functions models the path of the ball?

Ⓐ $y=3(x-5)^2+10$

Ⓑ $y=-3(x-5)^2+10$

Ⓒ $y=\frac{7}{25}(x-5)^2+10$

Ⓓ $y=-\frac{7}{25}(x-5)^2+10$

74. LEVEL:3

A train traveling at a speed of s miles per hour applies its brakes before a buffer stop. Assuming $d \geq 0$ and $s \geq 0$, the distance, d, in yards from the train to the buffer stop once the train comes to rest is:

$$d=0.5(-s^2-1.2s+184.6)$$

Which of the following equivalent expressions for d contains the traveling speed of the train, as a constant or coefficient, for which the train rests right at the buffer stop after applying its brakes?

Ⓐ $-0.5s^2-0.6s+92.3$

Ⓑ $-0.5(s-13)(s+14.2)$

Ⓒ $-0.5(s+0.6)^2+92.48$

Ⓓ $(6.5-0.5s)(s+14.2)$

75. LEVEL:3

A cereal company wants to enlarge the volume of the cylindrical container used for one of its products by enlarging the radius of the cylinder. The height must be $20cm$. The new volume of the cylinder is given by the equation

$$V(x) = 20\pi(5+x)^2$$

where x is the additional length of the radius in centimeters.
Which of the following equivalent expressions displays as a constant or coefficient the present value of the volume of the cylinder?

Ⓐ $\pi(20x^2 + 200x + 500)$

Ⓑ $20\pi(x^2 + 10x + 10) + 300\pi$

Ⓒ $20\pi(x^2 + 10x + 25)$

Ⓓ $20\pi x^2 + 200\pi x + 500\pi$

2.7 Interpreting nonlinear expressions : Exponential functions

1. Exponential functions

Exponential functions are functions in which the independent variable (usually denoted as 'x') appears as an exponent. They have the general form of

$$f(x) = a^x$$

where 'a' is the base of the exponential function. The base 'a' is a positive constant greater than 1, which determines the behavior of the function.

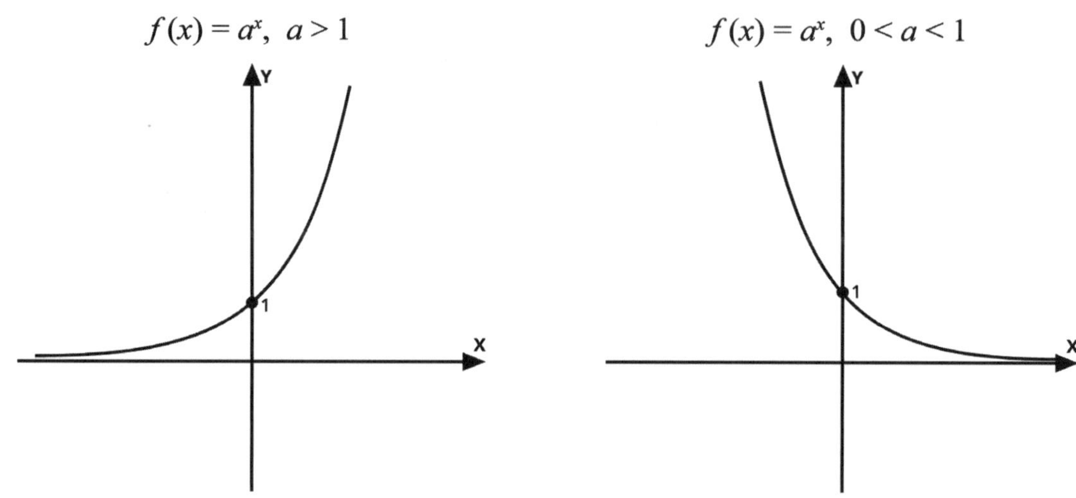

The entire graph lies above the x-axis, since the range of $y = a^x$ is all positive reals.

In summary, the properties of the graph of an exponential function $y = a^x$ are as follows:

- The graph passes through $(0, 1)$.
- When $a > 1$, the graph strictly increases as x, and is concave up.
- When $0 < a < 1$, the graph strictly decreases as x, and is concave up.
- The graph lies above the x-axis.

The graph has the

- x-axis as its horizontal asymptote.

2. Properties of exponential functions:

1. Growth or Decay: When 'a' is greater than 1, the exponential function represents growth. As 'x' increases, the function value increases rapidly. Conversely, when 'a' is between 0 and 1 (exclusive), the function represents decay. As 'x' increases, the function value decreases rapidly.

2. Exponential Growth/Decay Factor: The value of 'a' determines the growth or decay factor. For example, if 'a' is 2, the function will double each time 'x' increases by 1 (growth factor of 2). If 'a' is 0.5, the function will halve each time 'x' increases by 1 (decay factor of 0.5).

3. Horizontal Asymptote: An asymptote is a line that a function approaches but never crosses. Exponential functions have a horizontal asymptote at $y=0$ (the x-axis). As 'x' becomes very large (approaching positive or negative infinity), the function value gets arbitrarily close to 0.

4. Vertical Asymptote: Exponential functions do not have vertical asymptotes. They are defined for all real values of 'x'.

3. What is Compound Interest?

Compound interest is paid on the original principal and on the accumulated past interest.
If

- P is the principal (the initial amount you borrow or deposit)
- r is the annual rate of interest (percentage)
- n is the number of years the amount is deposited or borrowed for.
- A is the amount of money accumulated after n years, including interest.

When the interest is compounded once a year:

$$A = P(1+\frac{r}{100})^n$$

4. What if interest is paid more frequently for n years?

(1) Annually $= P\times(1+\frac{r}{100})^n$ (annual compounding)

(2) Quarterly $= P(1+\frac{r}{4\times 100})^{4n}$ (quarterly compounding)

(3) Monthly $= P(1+\frac{r}{12\times 100})^{12n}$ (monthly compounding)

76. LEVEL:1

The population (P) of a city is increasing at a rate of 2% per year. If the current population is $100{,}000$. write an equation to represent the population (P) after x years.

77. LEVEL:3

Reem is a test driver for an automobile company. The following formula gives the total distance, d, in feet that Reem drove a luxury car in the first t seconds after idling at a speed of 0 miles per hour, up to the time when she passed a particular safety cone.

$$d = 15.69\,t^2$$

Compared to the time it took Reem to pass the safety cone, how long did it take to pass a sensor that was $\frac{4}{9}$ of the distance from the start?

Ⓐ $\frac{16}{81}$ of the time it required to pass the cone

Ⓑ $\frac{4}{9}$ of the time it required to pass the cone

Ⓒ $\frac{2}{3}$ of the time it required to pass the cone

Ⓓ $\frac{9}{4}$ of the time it required to pass the cone

78.

LEVEL:3

Under ideal conditions, Lemna minor (common duckweed) is a fast-growing fern that can double its area every 2 days. Assume the growth is unrestricted, and that the duckweed initially covers 10 square centimeters (cm^2) in area. Which of the following functions, f, models the area (in cm^2) the duckweed covers after d days?

Ⓐ $f(d) = 10\,(0.5)^{\frac{d}{2}}$

Ⓑ $f(d) = 2 \times 10^d$

Ⓒ $f(d) = 10 \times 2^d$

Ⓓ $f(d) = 10 \times 2^{\frac{d}{2}}$

79.

LEVEL:4

Black tea is prepared by pouring boiling water (100 $°C$ onto tea leaves and allowing the tea to brew in a pot or cup. In a room whose temperature is 20 $°C$, the tea reaches a temperature of 60 $°C$ after about 4 minutes.

The temperature of the tea as a function of time can be modeled by an exponential function. Which of the following functions, T, best models the temperature of the cup of tea t minutes after pouring boiling water onto the leaves?

Ⓐ $T(t) = 20 + 40\,(0.84)^t$

Ⓑ $T(t) = 40 + 20\,(0.84)^t$

Ⓒ $T(t) = 40 \cdot (0.84)^t$

Ⓓ $T(t) = 20 \cdot (0.84)^t$

80.

LEVEL:4

The temperature T in degrees Celsius of a chilled drink after m minutes sitting on a table is given by the following function.

$$T(m)=32-28\cdot 3^{-0.05m}$$

What is the best interpretation of the number 32 in this function?

Ⓐ The drink is originally 32 degrees Celsius.

Ⓑ Every 32 minutes, the temperature warms by 3 degrees Celsius.

Ⓒ After 32 minutes, the drink will fully warm to the ambient temperature.

Ⓓ After sitting for a very long time, the drink will warm up to 32 degrees Celsius.

2.8 End Behavior of Polynomial Graph

1. Polynomials

Polynomials are algebraic expressions that consist of variables, coefficients, and exponents. They are built from addition, subtraction, and multiplication operations. The general form of a polynomial is given by:

$$P(x) = a_n x^n + a_{n-1} x^{n-1} + \dots + a_1 x + a_0$$

Key features of polynomials include:

- They only involve non-negative integer exponents.
- The coefficients (a_n, a_{n-1}, $\cdots$,a_1, a_0) are constants.
- They are defined for all real values of 'x'.
- The degree of a polynomial is the highest power of 'x' in the expression.
- The leading coefficient is the coefficient of the term with the highest degree.

Examples of polynomials :

$3x^2 - 5x + 2$

$4x^3 + 2x^2 - x + 7$

2. Non-polynomials

Non-polynomials, also known as transcendental functions, are mathematical expressions that cannot be expressed in the form of a polynomial. They involve other mathematical operations like exponential, logarithmic, trigonometric, or inverse trigonometric functions.

Examples of non-polynomials :

$\sqrt{x}$ (square root function)

$\sin(x)$(sine function)

e^x (exponential function)

$\ln(x)$ (natural logarithm function)

$\frac{1}{x}$ (reciprocal function)

3. Polynomial End Behavior

Every polynomial function either approaches infinity or negative infinity as x increases and decreases without bound. Which way the function goes as x increases and decreases without bound is called its end behavior.

(1) If the degree n of a polynomial is even, then the arms of the graph are either both up or both down.

(2) If the degree n is odd, then one arm of the graph is up and one is down.

(3) If the leading coefficient is positive, the right arm of the graph is up.

(4) If the leading coefficient is negative, the right arm of the graph is down.

4. Graphing Higher Degree Polynomials

IF $\alpha<\beta<\gamma<\delta$,

(1) $y=(x-\alpha)(x-\beta)(x-\gamma)$

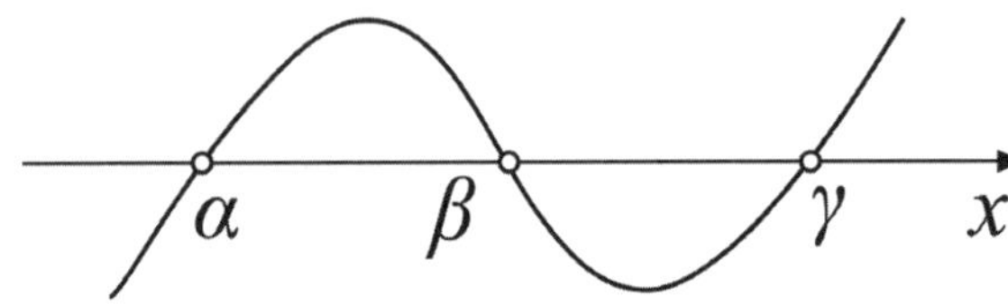

(2) $y=(x-\alpha)(x-\beta)(x-\gamma)(x-\delta)$

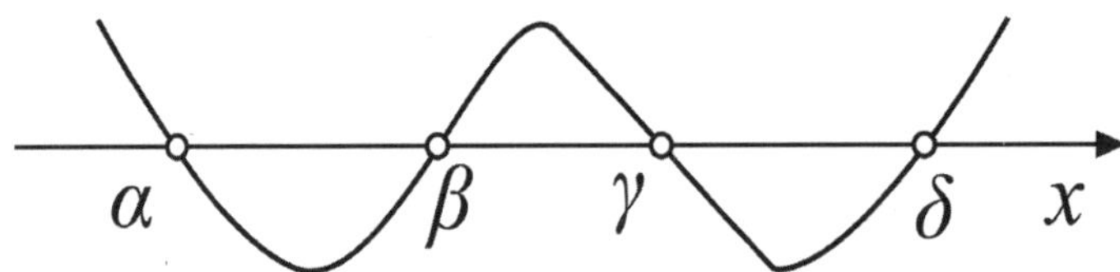

5. General Inequality and The Graph

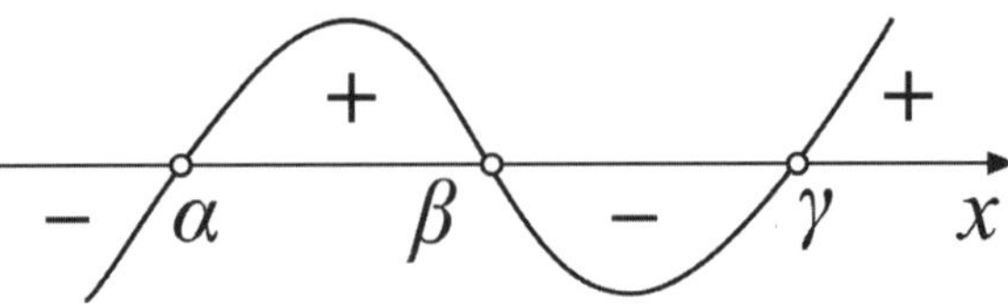

(1) $(x-\alpha)(x-\beta)(x-\gamma)>0 \Leftrightarrow \alpha<x<\beta,\ x>\gamma$

(2) $(x-\alpha)(x-\beta)(x-\gamma)<0 \Leftrightarrow x<\alpha,\ \beta<x<\gamma$

81. LEVEL:3

The function $p(t)$ is a polynomial of t such that $(t-10)$, $(22-t)$, $(t+10)$, and $(20+t)$ are all factors of $p(t)$. Which of the following could be the graph of $y=p(t)$ in the ty-plane?

Ⓐ

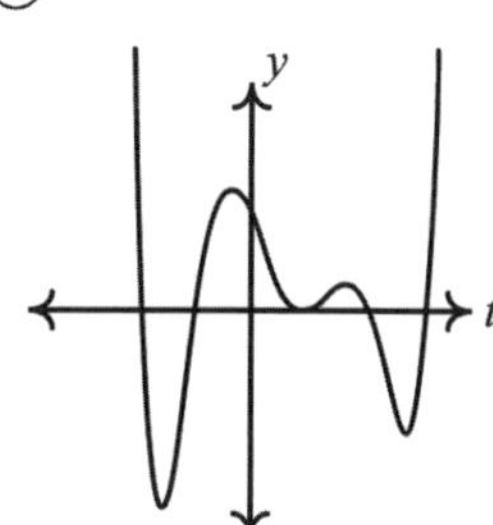

Ⓑ

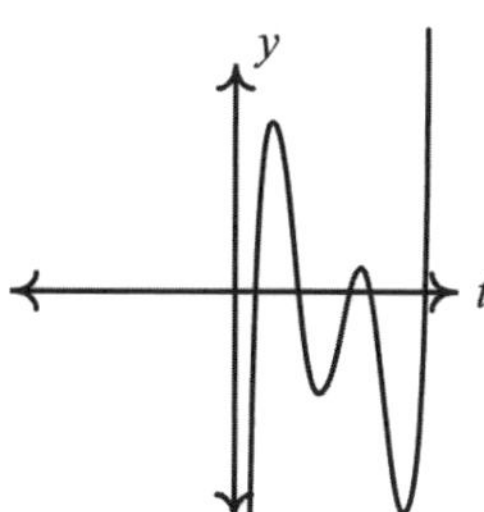

Ⓒ

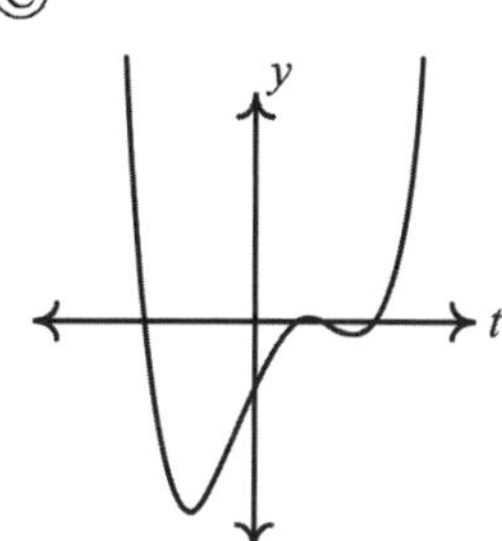

Ⓓ

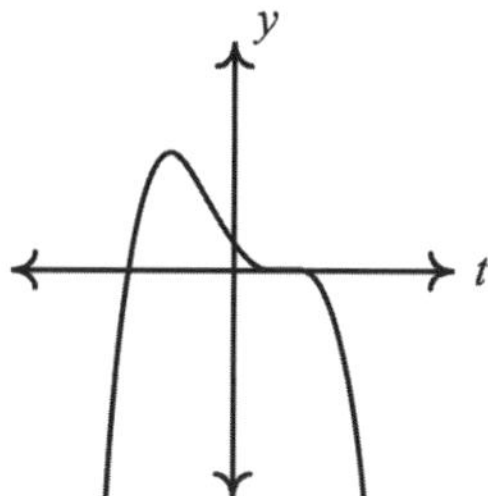

82. LEVEL:3

Which of the following graphs appears to represent a polynomial function with a double zero?

Ⓐ

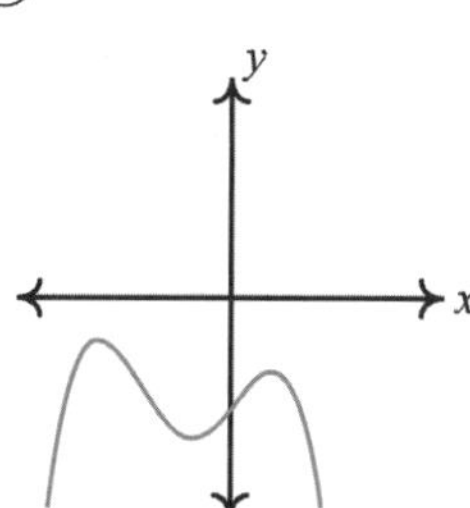

Ⓑ

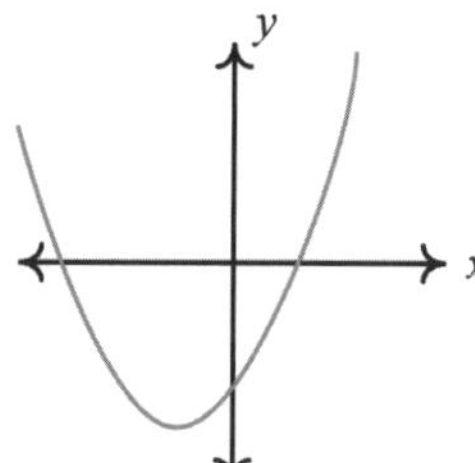

Ⓒ

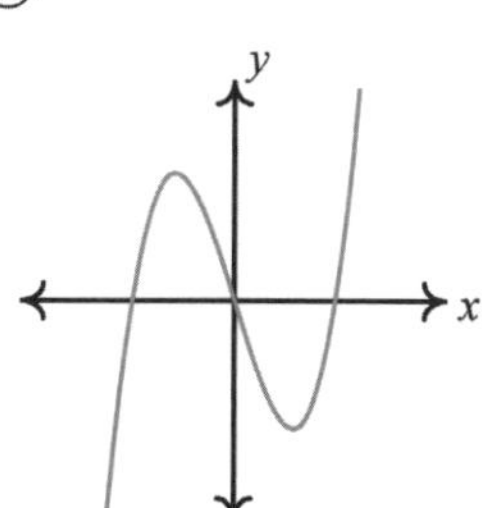

Ⓓ

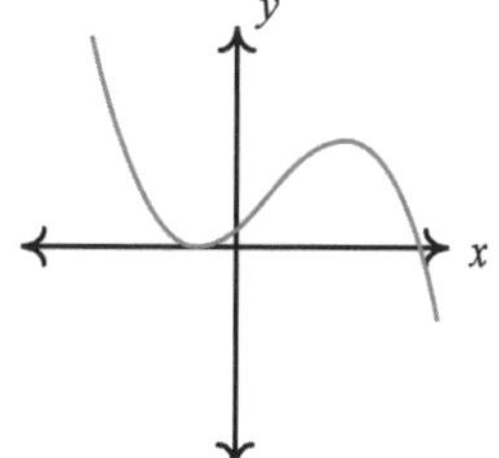

83. LEVEL:4

A function w is a defined as $w(x)=ax^2+bx+c$ where a, b, and c are constants. If $a=3$ and $w(3)=w(15)=0$, then what is the absolute value of b?

84. LEVEL:4

A function s is defined as $s(x)=(x-4)(x-5)^2$. A functionh is defined as $h(x)=(x-a)s(x)$. For some constant a, $(x-a)^3$ is a factor of h. What is $s(a)$?

0

2.9 Transforming Function

1. Translations of Graphs:

For $y=f(x)$,

(1) $y=f(x-a)$: Horizontally through a units

(2) $y=f(x)+b$: Vertically through b units

(3) $y=f(x-a)+b$: Horizontally through a units and vertically through b units

2. Transformations of Graphs : Reflections and Stretchs/Compressions

For $y=f(x)$,

(1) $y=-f(x)$: symmetric with respect to the x-axis

(2) $y=f(-x)$: symmetric with respect to the y-axis

(3) $y=-f(-x)$: symmetric with respect to the origin

(4) $y=pf(x)$: The effect of changes in p is to vertically stretch the graph of $y=f(x)$ by a factor of p

(5) $y=f(qx)$: The effect of changes in p is to horizontally stretch the graph of $y=f(x)$ by a factor of $\frac{1}{q}$.

3. Even and Odd functions

(1) Even Function

Let $f(x)$ be a real-valued function of a real variable. Then f is even if the following equation holds for all x in the domain of f : $f(x)=f(-x)$

Geometrically, the graph of an even function is symmetric with respect to the y-axis, meaning that its graph remains unchanged after reflection about the y-axis.

Example ▌ $y=x^2$, $y=x^4$, $y=\cos x$

(2) Odd Function

Let $f(x)$ be a real-valued function of a real variable. Then f is odd if the following equation holds for all x in the domain of f: $-f(x)=f(-x)$

Geometrically, the graph of an odd function has rotational symmetry with respect to the origin, meaning that its graph remains unchanged after rotation of 180 degrees about the origin.

Example ▌ $y=x$, $y=x^3$, $y=\sin x$

85.

LEVEL:2

Functions $p(x)$ and $w(x)$ are graphed in the xy-plane. The graph of $y=p(x)$ is equivalent to the graph of $y=w(x)$ translated 4 units upward and 3 units to the left, where the positive x-direction is to the right and the positive y-direction is upward.

Which of the following correctly relates $w(x)$ and $p(x)$?

Ⓐ $w(x)=p(x-3)+4$

Ⓑ $w(x)=p(x+3)+4$

Ⓒ $w(x)=p(x-3)-4$

Ⓓ $w(x)=p(x+3)-4$

86.

LEVEL:3

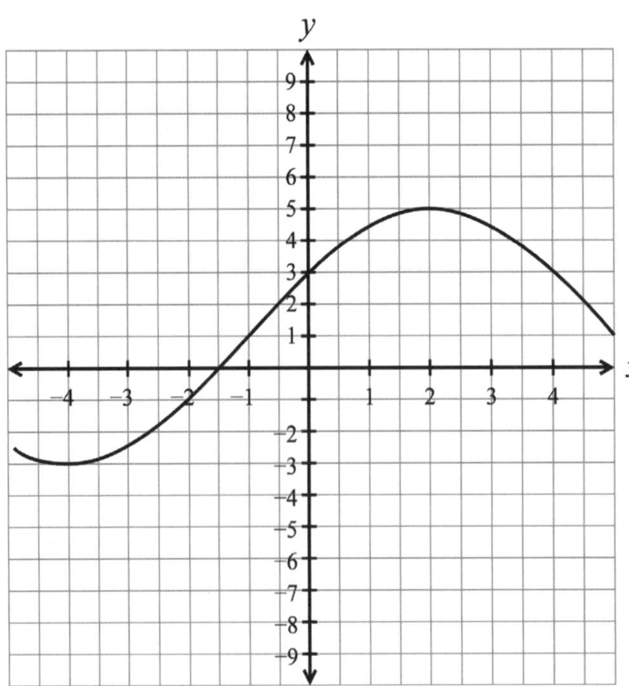

The graph of $y=-f(-x)$ is shown above. For which value of x is it true that $f(x)=1$?

87. LEVEL:3

The graph of function h is the graph of function g stretched vertically by a factor of 3 and reflected over the y-axis. Which of the following correctly defines function h?

Ⓐ $h(x) = -3g(x)$

Ⓑ $h(x) = 3g(-x)$

Ⓒ $h(x) = g(-3x)$

Ⓓ $h(x) = -g(3x)$

88. LEVEL:3

The graph of $9x - 10y = 19$ is translated down 4 units in the xy-plane. What is the x-coordinate of the x-intercept of the resulting graph?

2.10 The Graph of Rational Function

1. The Graph of $y = \dfrac{1}{x}$ And Its Transformation

Rational functions are characterised by the presence of both a horizontal asymptote and a vertical asymptote.

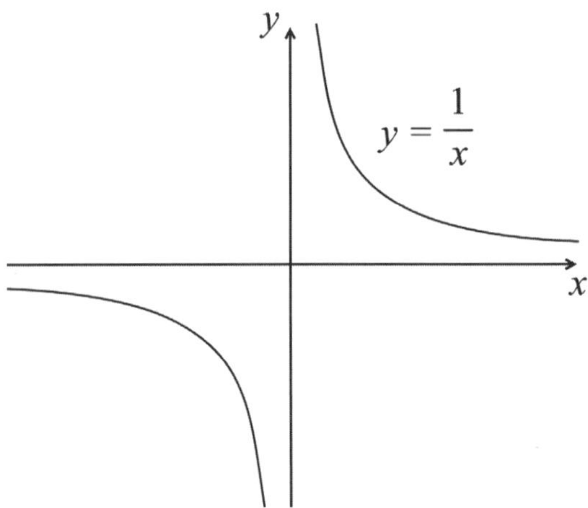

Any graph of a rational function can be obtained from the reciprocal function $f(x) = \dfrac{1}{x}$ by a combination of transformations, including a translation, stretches and compressions.

Type 1: Vertical Compression

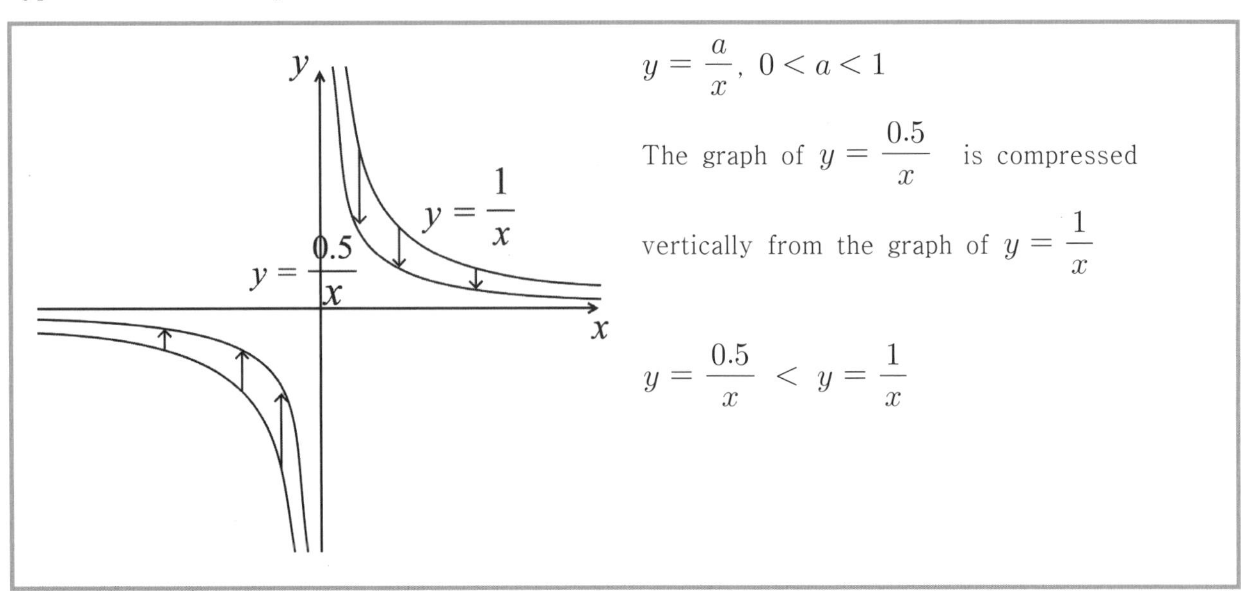

$y = \dfrac{a}{x}$, $0 < a < 1$

The graph of $y = \dfrac{0.5}{x}$ is compressed vertically from the graph of $y = \dfrac{1}{x}$

$y = \dfrac{0.5}{x} < y = \dfrac{1}{x}$

Type 2: Vertical Stretch

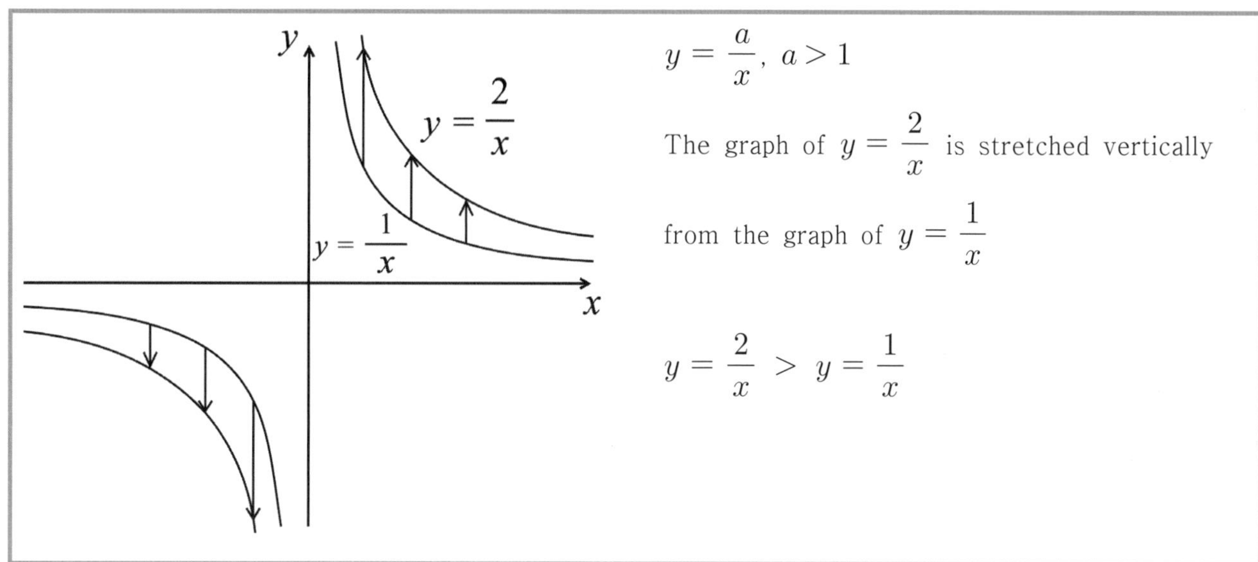

$y = \frac{a}{x}$, $a > 1$

The graph of $y = \frac{2}{x}$ is stretched vertically from the graph of $y = \frac{1}{x}$

$y = \frac{2}{x} > y = \frac{1}{x}$

Type 3: Horizontal Stretch

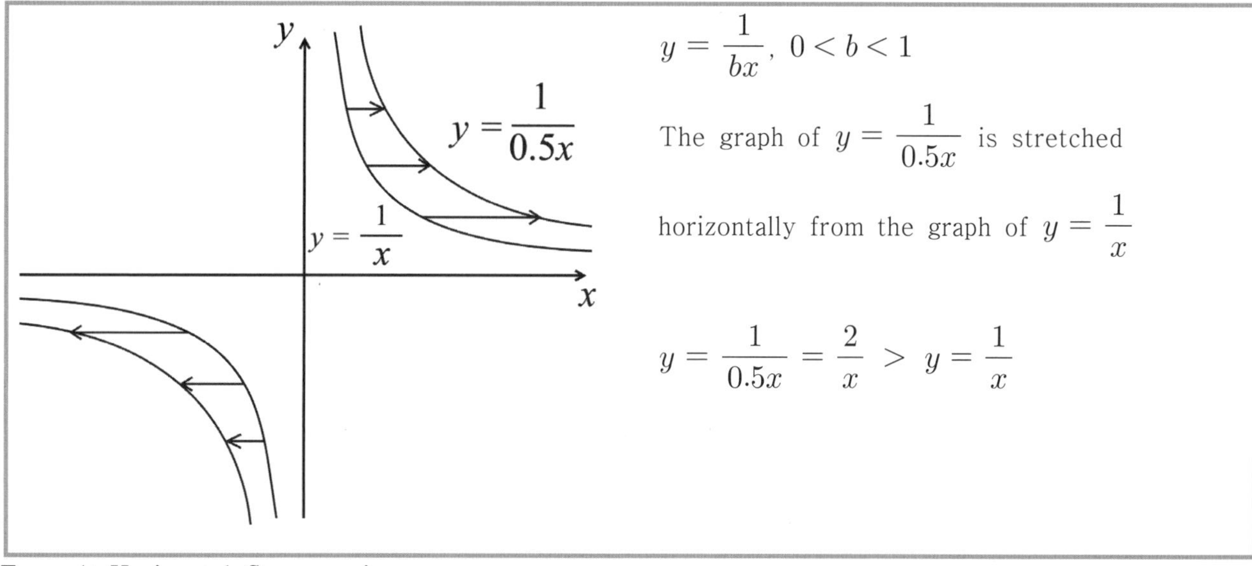

$y = \frac{1}{bx}$, $0 < b < 1$

The graph of $y = \frac{1}{0.5x}$ is stretched horizontally from the graph of $y = \frac{1}{x}$

$y = \frac{1}{0.5x} = \frac{2}{x} > y = \frac{1}{x}$

Type 4: Horizontal Compression

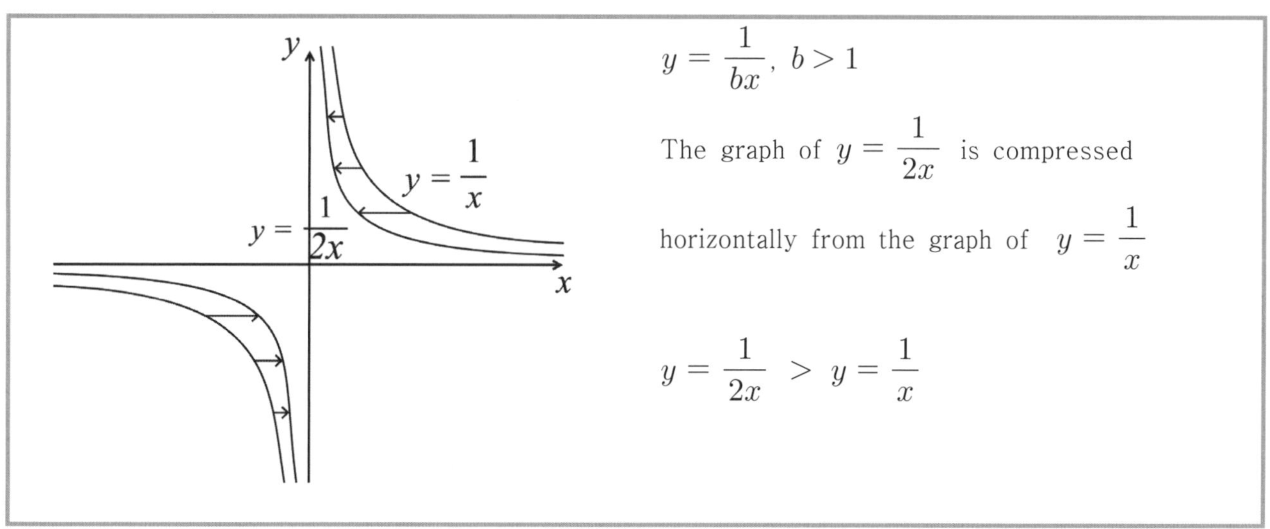

$y = \frac{1}{bx}$, $b > 1$

The graph of $y = \frac{1}{2x}$ is compressed horizontally from the graph of $y = \frac{1}{x}$

$y = \frac{1}{2x} > y = \frac{1}{x}$

Type 5: Horizontal Translation to the Right

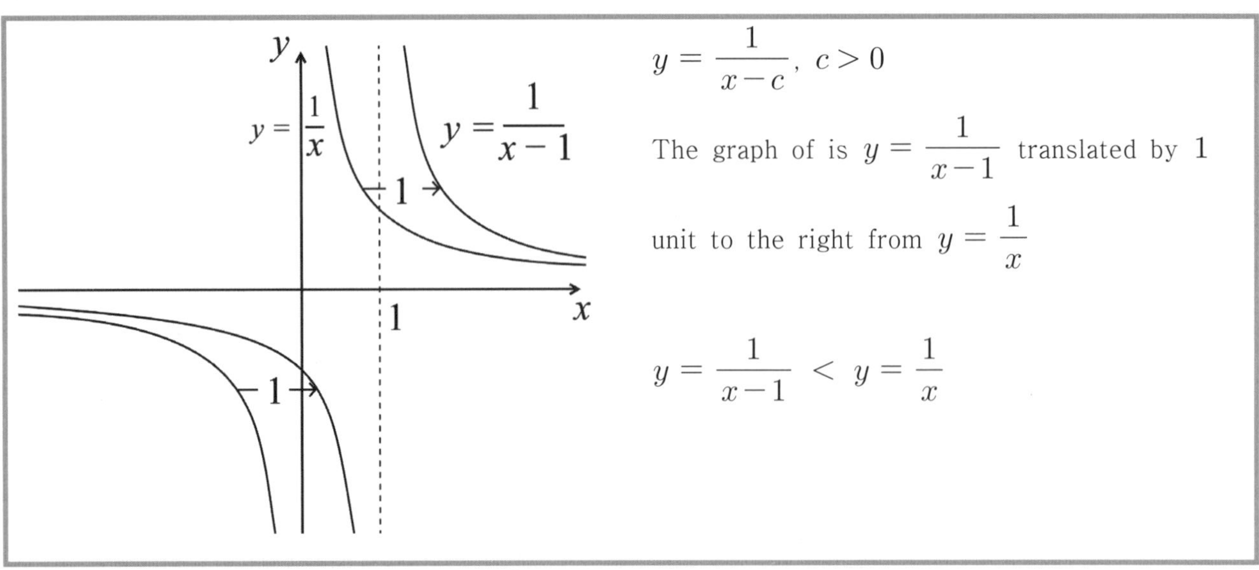

$y = \dfrac{1}{x-c},\ c > 0$

The graph of is $y = \dfrac{1}{x-1}$ translated by 1 unit to the right from $y = \dfrac{1}{x}$

$y = \dfrac{1}{x-1} < y = \dfrac{1}{x}$

Type 6: Horizontal Translation to the Left

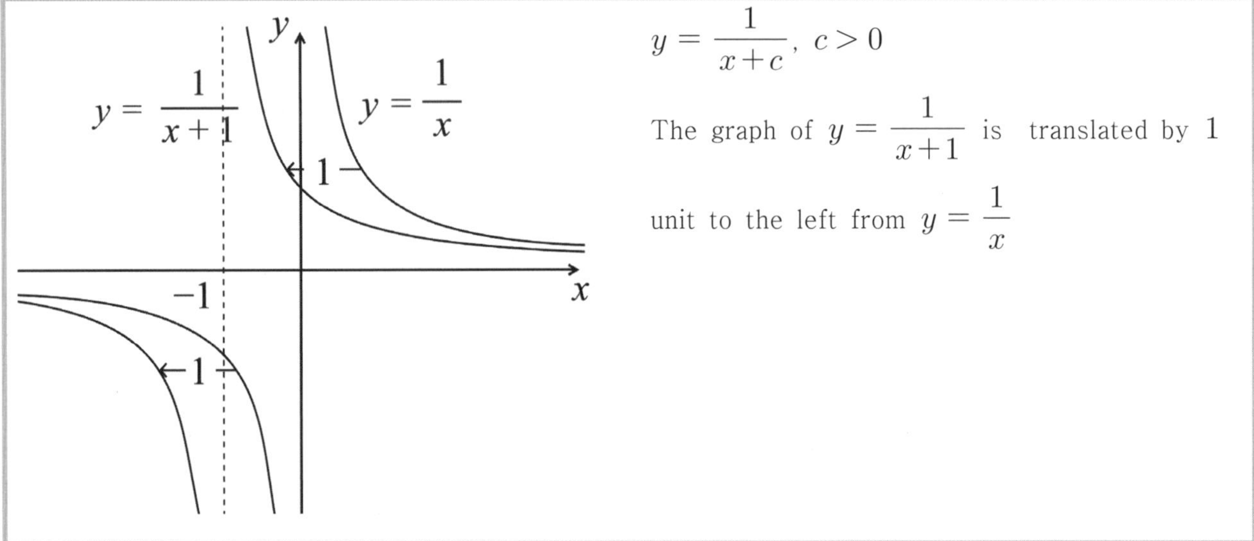

$y = \dfrac{1}{x+c},\ c > 0$

The graph of $y = \dfrac{1}{x+1}$ is translated by 1 unit to the left from $y = \dfrac{1}{x}$

Type 7: Vertical Translation Upwards

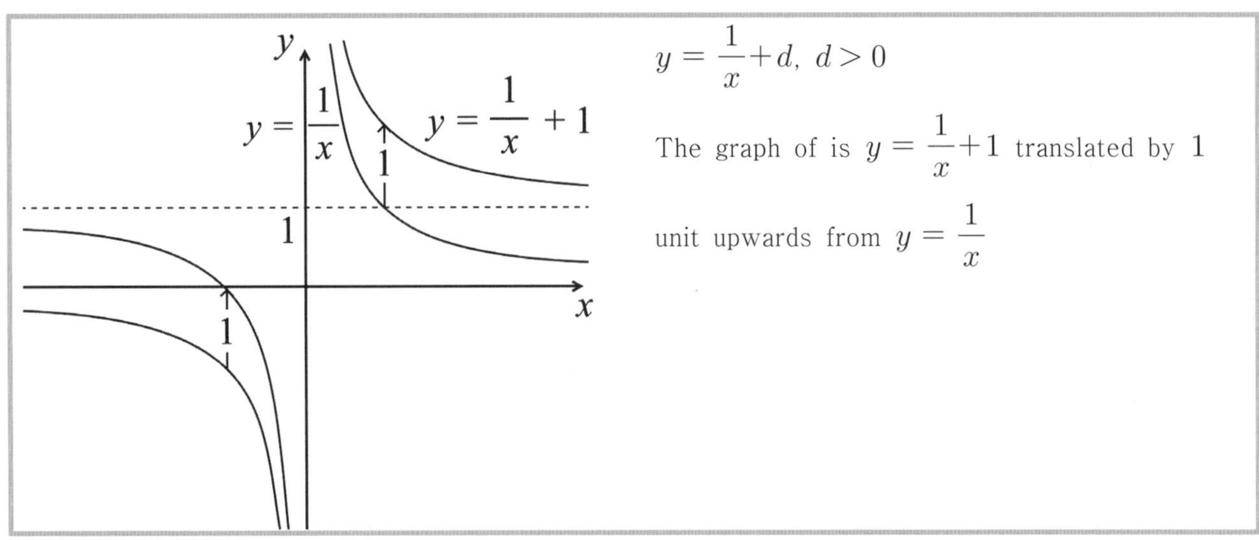

$y = \dfrac{1}{x}+d,\ d > 0$

The graph of is $y = \dfrac{1}{x}+1$ translated by 1 unit upwards from $y = \dfrac{1}{x}$

Type 8: Vertical Translation Downwards

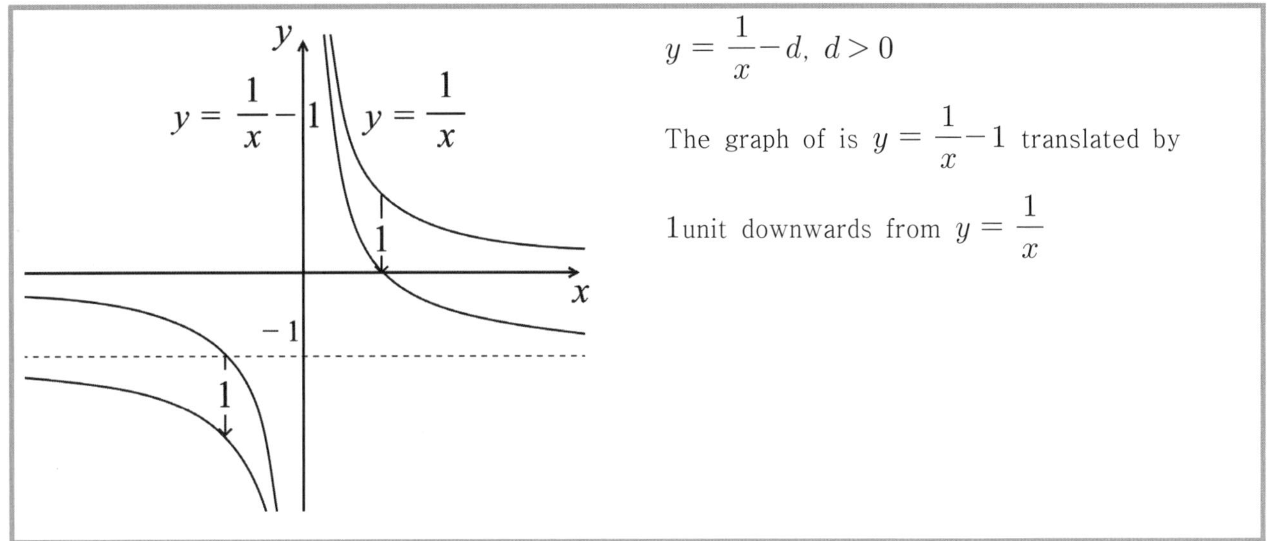

$y=\frac{1}{x}-d,\ d>0$

The graph of is $y=\frac{1}{x}-1$ translated by 1unit downwards from $y=\frac{1}{x}$

2. The Graph of $y=\frac{ax+b}{cx+d}$

In the case of the rational function

$$y=\frac{ax+b}{cx+d}$$

The numerator is a linear polynomial (degree 1) and the denominator is also a linear polynomial. Let's explore its properties:

- Domain : The rational function is defined for all real values of 'x' except when the denominator $cx+d$ is equal to zero. So, the function is undefined when $cx+d=0$, or when $x=-\frac{d}{c}$. This value is called a vertical asymptote of the function.

- Vertical Asymptotes : If $c\neq 0$, the vertical asymptote of the function occurs at $x=-\frac{d}{c}$. The graph of the function approaches the vertical line $x=-\frac{d}{c}$ as 'x' approaches positive or negative infinity. The vertical asymptote represents the values that 'x' cannot take for the function to be defined.

- Horizontal Asymptotes : If the degrees of numerator and denominator of rational functions are equal (i.e., degree of $ax+b$ = degree of $cx+d$), the horizontal asymptote is the ratio of the leading coefficients $\frac{a}{c}$. The graph of the function approaches the horizontal line $y=\frac{a}{c}$.

- Intercepts : The x-intercepts occur when the numerator $(ax+b)$ equals zero, so giving $x=-\frac{b}{a}$.

The y-intercept occurs when 'x' is zero, giving $y=\frac{b}{d}$.

89. LEVEL:2

A rational equation is graphed in the xy-plane. The graph has a vertical asymptote with the equation $x=3$ and a horizontal asymptote with the equation $y=0$. Which of the following could be the equation?

Ⓐ $y=\dfrac{1}{x+3}$

Ⓑ $y=\dfrac{1}{x-3}$

Ⓒ $y=\dfrac{1}{x}+3$

Ⓓ $y=\dfrac{1}{x}-3$

90. LEVEL:3

The equations $y=\dfrac{3}{x+4}$ and $y=\dfrac{3}{x+4}-3$ are graphed in the xy-plane. Which of the following must be true of the asymptotes of the graphs of the two equations?

Ⓐ Both graphs have a vertical asymptote at $x=4$.

Ⓑ Both graphs have a vertical asymptote at $x=-4$.

Ⓒ $y=\dfrac{3}{x+4}$ has a horizontal asymptote at $y=0$, and $y=\dfrac{3}{x+4}-3$ has a horizontal asymptote at $y=3$.

Ⓓ $y=\dfrac{3}{x+4}$ has a vertical asymptote at $x=0$, and $y=\dfrac{3}{x+4}-3$ has a vertical asymptote at $x=-3$.

91. LEVEL:4

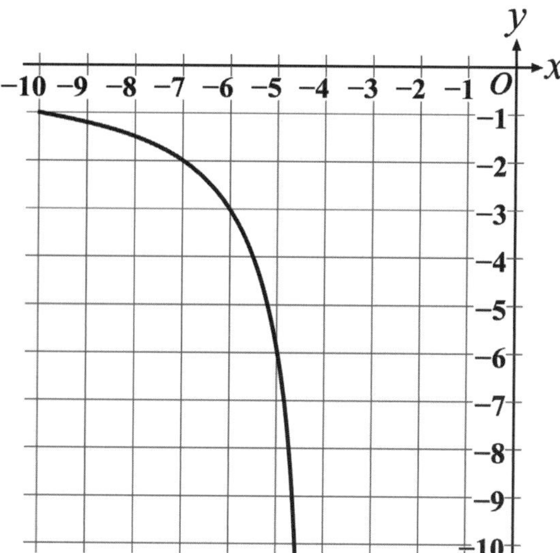

The rational function f is defined by an equation in the form $f(x)=\dfrac{a}{x+b}$, where a and b are constants. The partial graph of $y=f(x)$ is shown. If $g(x)=f(x+4)$, which equation could define function g?

Ⓐ $g(x)=\dfrac{6}{x}$

Ⓑ $g(x)=\dfrac{6}{x+4}$

Ⓒ $g(x)=\dfrac{6}{x+8}$

Ⓓ $g(x)=\dfrac{6(x+4)}{x+4}$

2.11 The Graph of Irrational Function of Transformations on $y=\sqrt{x}$

1. The basic square root function, $y=\sqrt{x}$

The basic square root function is given by $y=\sqrt{x}$

The curve of the graph of $y=\sqrt{x}$ is one that begins at the origin and rises in a concave downward way, so that the rate of rise is forever slowing although never stopping to rise.

x	0	1	4	9	16	25
$y=\sqrt{x}$	0	1	2	3	4	5

2. Transformations on $y=\sqrt{x}$

Parent Function	$y=\sqrt{x}$
Vertical Translation by k	$y=\sqrt{x}+k$
Horizontal Translation by h	$y=\sqrt{x-h}$
Vertical Translation by k and Horizontal Translation by h	$y=\sqrt{x-h}+k$
Reflection in the $x-$axis	$y=-\sqrt{x}$
Stretch by factor of a	$y=a\sqrt{x}$
Stretch by factor of a with Reflection in the $x-$axis	$y=-a\sqrt{x}$

92.

LEVEL:2

The function $f(x)=\sqrt{x-7}$ is graphed in the xy-plane as $y=f(x)$. Which of the following statements about the function's graph is true?

Ⓐ The graph of the function is always decreasing.

Ⓑ The graph of the function intersects the x-axis at $(7,0)$.

Ⓒ The graph of the function intersects the y-axis at $(0,-7)$.

Ⓓ The graph of the function is symmetric about the line $x=7$.

93.

LEVEL:3

Graph the function $y=-\sqrt{x+3}-5$.

2.12 Isolating Quantities

Isolating quantities is an essential skill in SAT Math, especially when solving algebraic equations or manipulating formulas. It involves rearranging an equation or formula to express a specific variable or quantity in terms of the other variables. This process allows you to isolate the desired quantity and solve for its value. Here's a step-by-step approach to isolating quantities:

1. Identify the desired quantity: Determine which variable or term you want to isolate and solve for.

2. Simplify the equation or formula: Simplify the given equation or formula by combining like terms, distributing, or performing any necessary operations.

3. Use inverse operations: Apply inverse operations to isolate the desired quantity. Inverse operations are operations that "undo" each other. For example:

- To isolate a term with addition, subtract the same value from both sides of the equation.
- To isolate a term with subtraction, add the same value to both sides of the equation.
- To isolate a term multiplied by a constant, divide both sides of the equation by that constant.
- To isolate a term divided by a constant, multiply both sides of the equation by that constant.

Example ▌ Solve for x : $\frac{y}{x+3} = 2\ (x \neq -3)$

Solution ▌ $2(x+3) = y \Rightarrow 2x + 6 = y \Rightarrow 2x = y - 6$

$$\therefore x = \frac{y-6}{2}$$

94. LEVEL:2

$$c = \frac{ab}{a+b}$$

For two capacitors, wired in series, the equivalent capacitance, c, can be expressed in terms of the capacitance, a and b, of each capacitor, with the above equation. Which of the following correctly expresses the capacitance of capacitor a in terms of capacitor b and the equivalent capacitance c?

Ⓐ $a = \frac{c}{b-c}$

Ⓑ $a = cb(b-c)$

Ⓒ $a = \frac{cb}{b-c}$

Ⓓ $a = \frac{cb}{b+c}$

95. LEVEL:4

$$x = v_0 t + \frac{1}{2}at^2$$

The horizontal displacement, x, of an object with constant acceleration, a, initial velocity, v_0, at elapsed time, t, is given by the above equation. Which of the following equations correctly shows the acceleration in terms of displacement, initial velocity, and time?

Ⓐ $a = \frac{2\sqrt{x - tv_o}}{t}$

Ⓑ $a = \frac{2(x - v_o)}{t}$

Ⓒ $a = \frac{2(x - tv_o)}{t^2}$

Ⓓ $a = \frac{2x}{t^3 v_o}$

2.13 Ratios, Rates and Proportions

1. Ratio:

Ratios are used to compare two or more quantities. They express the relationship between the quantities in terms of division. Ratios can be written in different forms, such as $\frac{a}{b}$, a to b, or $a : b$.

Example ▌

- The ratio of boys to girls in a class is $3 : 5$. This means that for every 3 boys, there are 5 girls.
- The ratio of the lengths of two sides of a rectangle is $2 : 5$. This means that the length of one side is $\frac{2}{5}$ times the length of the other side.

When working with ratios on the SAT, you may be asked to simplify ratios, find missing values in a ratio, or use ratios to solve problems involving proportions or comparisons.

2. Rates:

Rates involve a ratio between two different units of measurement. They express how one quantity changes with respect to another quantity. Rates are often written as a quantity per unit, such as miles per hour, dollars per hour, or items per minute.

Example ▌

- The car is traveling at a rate of 60 miles per hour. This means that for every hour, the car travels a distance of 60 miles.
- A worker earns $\$12$ per hour. This means that for each hour worked, the worker earns $\$12$.

On the SAT, rate problems may involve calculating average rates, finding missing values in a rate, or using rates to solve problems involving distance, time, or cost.

Average Rate Formula

$$\text{Average Rate} = \frac{\text{Total Distance}}{\text{Total Time}}$$

3. Proportions:

Proportions are statements that two ratios or rates are equal. Proportions are useful for solving problems that involve unknown values. Proportions can be solved using cross−multiplication or equivalent fractions.

Example ▮

- If 4 men can paint a house in 6 days, how many men are needed to paint the house in 3 days? This problem can be solved using a proportion, 4 men : 6 days = x men : 3 days. By multiplying and solving for 'x', you can find the number of men needed.

4. Variation or Proportionality Formulas

(1) Direct : $y = kx$

(2) Inverse : $y = \dfrac{k}{x}$

(3) Joint : $z = kxy$

(4) Inverse Joint : $z = \dfrac{kx}{y}$

(5) Direct to power : $y = kx^n$

(6) Inverse to Power : $y = \dfrac{k}{x^n}$

96. LEVEL:3

During a timed test, Alexander typed 742 words in 14 minutes. Assuming Alexander works at this rate for the next one hour, which of the following best approximates the number of words he would type in that hour?

Ⓐ 53

Ⓑ 840

Ⓒ 3,180

Ⓓ 44,520

97. LEVEL:3

Elena is conducting a study about the effects of toxins in the water on the hormones of fish. Elena surveys 350 male fish in a river and finds that 150 of the male fish have egg cells growing inside them. According to Elena's survey, what is the ratio of male fish with egg cells to male fish without egg cells in the river?

Ⓐ $3:4$

Ⓑ $3:7$

Ⓒ $4:5$

Ⓓ $4:7$

98.

LEVEL:3

Alan drives an average of 100 miles each week. His car can travel an average of 25 miles per gallon of gasoline. Alan would like to reduce his weekly expenditure on gasoline by $\$5$. Assuming gasoline costs $\$4$ per gallon, which equation can Alan use to determine how may fewer average miles, m, he should drive each week?

Ⓐ $\frac{25}{4}m = 95$

Ⓑ $\frac{25}{4}m = 5$

Ⓒ $\frac{4}{25}m = 95$

Ⓓ $\frac{4}{25}m = 5$

2.14 Percents

The concept of percents is a fundamental topic in SAT Math and involves understanding and working with proportions and ratios expressed as fractions of 100. Percentages are used to express parts of a whole or compare quantities. Here's an overview of key concepts related to percents on the SAT Math.

1. Definition of Percent :

A percent is a ratio or fraction that compares a quantity to 100. The symbol "%" represents a percent. For example, 25% is equivalent to the fraction $\frac{25}{100}$ or the decimal 0.25.

Definition of Percent

$$0 \le \text{Percent } (\%) = \frac{\textit{Partial Amount}}{} \le 100$$

2. Converting Between Percents, Fractions, and Decimals

- To convert a percent to a decimal, divide the percent by 100. For example, 50% is equivalent to 0.50.
- To convert a fraction or decimal to a percent, multiply the fraction or decimal by 100 and add the "%" symbol. For example, 0.6 can be written as 60%, and $\frac{3}{5}$ can be written as 60%.

3. Percent Increase or Decrease

Percent increase or decrease is used to describe changes in quantity compared to the original value. To calculate a percent increase or decrease:

- Determine the amount of change (new value − original value).
- Divide the amount of change by the original value.
- Multiply the result by 100 to get the percent increase or decrease.

$$\text{Percent Increase } (\%) = \frac{New - Original}{Original} \times 100$$

$$\text{Percent decrease } (\%) = \frac{Original - New}{Original} \times 100$$

99. LEVEL:3

The chef has 36 pounds of strip loin steak. The amount of weight lost from the trim on each pound is 35%, and 75% more of the steak's weight is lost after cooking it. How many pounds of trimmed, cooked strip loin will the chef have left to serve to the customers?

Ⓐ 5.85
Ⓑ 12.6
Ⓒ 17.55
Ⓓ 23.4

100. LEVEL:3

Ivy is downloading a computer program from the Internet. After 8 minutes, the computer program is 35% downloaded. If the computer program continues to download at the current rate, about how much longer will it take for Ivy's computer to finish downloading the program?

Ⓐ 12 minutes
Ⓑ 15 minutes
Ⓒ 18 minutes
Ⓓ 23 minutes

101. LEVEL:4

Baby name	Denis	Dimitri	Lea	Tanya
Frequency	13	27	125	400

The table above displays the number of babies per million babies born in 1985 with each of 4 names. In 1985, about what percentage of babies were named Lea or Dimitri?

Ⓐ 0.000152%
Ⓑ 0.0152%
Ⓒ 0.98%
Ⓓ 1.52%

102. LEVEL:4

A high school's graduation rate is defined to be the percentage of the senior class that graduates. Last year 406 of Sagamore High School's 452 seniors graduated. This year the school expects the previous year's graduation rate to increase by approximately 2 percentage points. If there are 436 students in this year's senior class, which of the following best approximates the number of seniors that Sagamore High School expects to graduate this year?

Ⓐ 390 students
Ⓑ 400 students
Ⓒ 410 students
Ⓓ 420 students

103. LEVEL:4

How many liters of a 25% saline solution must be added to 3 liters of a 10% saline solution to obtain a 15% saline solution?

2.15 Units

1. Converting units

Converting units is a common topic in SAT Math, and it involves changing measurements from one unit to another. It's important to be able to convert between different units of measurement, such as length, area, volume, weight, and time. Here's an overview of the concept of converting units and how to approach related problems:

■ Understanding Conversion Factors :
Conversion factors are ratios that relate different units of measurement. These ratios allow you to convert from one unit to another. For example, 1 meter is equal to 100 centimeters, so the conversion factor is 1 meter/ 100 centimeters or $1m\ /\ 100cm$.

■ Setting Up Proportions :
To convert units, you can set up a proportion using the conversion factors. The units you want to convert from should be on opposite sides of the proportion, so they cancel out when you cross-multiply. For example, to convert 200 centimeters to meters using the conversion factor mentioned earlier, you would set up the proportion : $1m/100cm = x$ meters $/\ 200cm$.

■ Applying the Conversion :
Once you've set up the proportion, cross-multiply and solve for the desired unit.
In the example above, you would have : $1m \times 200cm = 100cm \times x\ m$. Simplifying, you find that 200 meters = $100x$, and dividing both sides by 100 gives $x = 2$ meters.

■ Pay Attention to Units:
When converting units, it's important to pay attention to the units in the problem and ensure that the units cancel out correctly during the conversion. This helps ensure that the final answer has the correct units.

■ Compound Conversions :
In some cases, you may need to perform compound conversions where you convert from one unit to another, and then to another unit. The same principles of setting up proportions and using conversion factors apply.

2. Conversion of Units

1) Price unit

1 penny = 1 cent

1 nickle = 5 cents

1 dime = 10 cents

1 quarter = 25 cents

1 dollar = 100 cents

1 dollar = 4 quarters = 10 dimes = 20 nickles = 100 cents

2) Volume

1 gallon = 4 quart = 8 pt (1 quart = 2 pint)

3) Weight

1 ton = 2000 Pounds (lb) , 1 lb = 16 Ounce(oz)

4) Length

1 yard = 3 feet, 1 foot = 12 inch

1 mile = 5,280 feet

1 mile = 1,760 yards

5) Density

$$Density = \frac{mass}{volume}$$

Note: Usually, the criteria for most unit conversions are given in the promblem on SAT Math.

3. Convert Units of Area and Volume

The objective of the lesson is to convert units of measure between dimensions including area and volume.

1 yard (yd) = 3 feet (ft)

1 mile = 1,760 yards

1 feet (ft) = 12 inches (in)

1 meter (m) = 100 centimeters (cm)

1 centimeters (cm) = 10 milimeters (mm)

1) Common conversions for Square Units

The table gives several common measurement conversions for square units.

Customary Unit	Metric Units
$1\,yd^2 = 9\,ft^2$	$1\,m^2 = 10{,}000\,cm^2$
$1\,ft^2 = 144\,in^2$	$1\,cm^2 = 100\,mm^2$

2) Common conversions for cubic units

The table gives several common measurement conversions for cubic units.

Customary Unit	Metric Units
$1\,yd^3 = 27\,ft^3$	$1\,m^3 = 1{,}000{,}000\,cm^3$
$1\,ft^3 = 1{,}728\,in^3$	$1\,cm^3 = 1{,}000\,mm^3$

104. LEVEL:2

Laurin is packing the books in her library. She's able to fit 2 paperback books in a box with no space left over, and 16 paperback books in a larger box with no space left over. If the dimensions of the first box are x by x by x, and the dimensions of the second box are kx by kx by kx, what is the value of k?

105. LEVEL:2

A sample of oak has a density of 807 kilograms per cubic meter. The sample is in the shape of a cube, where each edge has a length of 0.90 meters. To the nearst whole number, what is the mass, in the kilograms, of this sample?

106.
LEVEL:3

A certain park has an area of 11,863,808 square yards. What is the area, in square miles, of this park?

(Note: 1 mile = 1,760 yards)

Ⓐ 1.96

Ⓑ 3.83

Ⓒ 3,444.39

Ⓓ 6740.8

107.
LEVEL:3

A certain town has an area of 4.36 square miles. What is the area, in <u>square yards</u>, of this town?

(Note: 1 mile = 1,760 yards)

Ⓐ 404

Ⓑ 7,674

Ⓒ 710,459

Ⓓ 13,505,536

108.

LEVEL:4

A generator produces 6.5×10^2 kilojoules per centisecond $\left(\frac{kj}{cs}\right)$. A watt is equivalent to 1 joule per second $\left(\frac{j}{s}\right)$. What is the measured power of the generator in watts?

Ⓐ 6.5×10^7 watts

Ⓑ 6.5×10^2 watts

Ⓒ 6.5×10^1 watts

Ⓓ 6.5×10^3 watts

109.

LEVEL:4

The apparent brightness of a surface, in "lux", is found by measuring the energy of the light coming from the source and dividing by the area of the surface. Lux are equivalent to candelas per square meter. Abigail is building a new type of television screen and measures the brightness of glare to be 0.0002 kilocandelas per square centimeter. What is the brightness of glare expressed in lux?

Ⓐ 0.00002 lux

Ⓑ 0.002 lux

Ⓒ 20 lux

Ⓓ 2,000 lux

110. LEVEL:5

In 2012, an 11−year−old cheetah set a new record by running 100 meters in 5.95 seconds. During this record−breaking run, at what approximate speed was the cheetah traveling in miles per hour? (Note: There are 1.6 kilometers in 1 mile.)

Ⓐ 16.81 miles per hour
Ⓑ 34.27 miles per hour
Ⓒ 37.82 miles per hour
Ⓓ 60.50 miles per hour

Problem Solving and Data Analysis

A surprising proportion of mathematicans are accomplished
musicians. It is because music and mathematics
share patterns that are beautiful?
-MARTIN GARDNER (1914 - 2010)

☑ Problem Solving and Data Analysis section of the SAT Math

Problem Solving and Data Analysis is one of the major content areas covered in the SAT Math section. It assesses your ability to apply mathematical concepts and skills to real-world situations, analyze data, and solve problems using quantitative reasoning. Here are the key components of Problem Solving and Data Analysis on the SAT Math:

1. Interpreting and Analyzing Data :

You will encounter questions that require interpreting and analyzing data presented in various formats, such as graphs, tables, charts, and scatterplots. You'll need to understand the information provided and extract relevant details to answer the questions accurately.

2. Descriptive Statistics and Data Representation :

This includes questions related to mean, median, mode, range, standard deviation, and other measures of central tendency and variability. You'll need to analyze data sets, interpret statistical information, and draw conclusions based on the data provided.

3. Probability and Statistics :

Questions in this area require understanding basic probability concepts, such as probability calculations, combinations, permutations, and conditional probability. You'll need to apply probability principles to solve problems involving outcomes, events, and experiments.

4. Experiments and Studies :

These questions involve understanding the design and interpretation of experiments and studies. You may need to analyze experimental setups, identify control groups, recognize biases, evaluate sample sizes, and draw conclusions based on the results presented.

3.1 The Types of Data and Probability

In the context of SAT Math, the types of data refer to the different ways in which data can be classified and represented. Understanding the types of data is essential for interpreting and analyzing information presented in graphs, tables, and other data representations on the SAT.

1. Categorical Data:

Categorical data consists of distinct categories or groups. It is qualitative rather than quantitative. Examples include the type of car (sedan, SUV, truck), favorite color (red, blue, green), or grade level (freshman, sophomore, junior). Categorical data is often represented using bar graphs or pie charts.

2. Numerical Data:

Numerical data, also known as quantitative data, consists of numerical values that represent quantities or measurements. Numerical data can be further classified into the following subtypes:

1) Discrete Data:

Discrete data consists of individual, separate values that are typically whole numbers and have distinct boundaries. Examples include the number of siblings, the number of cars in a parking lot, or the number of pets in a household.

2) Continuous Data:

Continuous data consists of values that can take on any real number within a given range. It is often measured on a continuous scale. Examples include height, weight, temperature, or time. Continuous data is often represented using line graphs or histograms.

3) Univariate Data:

Univariate data refers to a data set that consists of a single variable or characteristic. It involves the analysis of a single set of observations or measurements. For example, if you collect data on the heights of a group of individuals, you would have univariate data. Univariate data analysis involves summarizing and describing the distribution, central tendency, and variability of the data using measures such as mean, median, mode, range, and standard deviation. Graphical representations such as histograms, box plots, or stem-and-leaf plots are commonly used to display univariate data.

4) Bivariate Data:

Bivariate data refers to a data set that involves two variables or characteristics. It consists of pairs of observations or measurements taken from two different variables. For example, if you collect data on the hours of study and test scores of a group of students, you would have bivariate data. Bivariate data analysis focuses on examining the relationship between the two variables and identifying patterns or trends. Common techniques for analyzing bivariate data include scatter plots, correlation coefficients, and regression analysis.

3. Probability

The College Board tests basic probability concepts on the SAT.

1) Probability indicates the likelihood that a specific event will occur. The probability of something happening can be expressed using any number from 0 to 1.

2) A probability of 0 means the event will NEVER happen. A probability of 1 indicates that an event will always happen.

3) The probability of an occurrence can be expressed by a simple formula:

$$\text{Probability} = \frac{\text{number of favorable outcomes}}{\text{number of possible outcomes}}$$

4) The probability of something not occurring is 1 minus the probability that it will occur:

$$\text{Probability of an event } \textit{not} \text{ occurring} = 1 - \frac{\text{number of favorable outcomes}}{\text{number of possible outcomes}}$$

4. Conditional Probability

Conditional probability is a measure of the probability of an event occurring given that another event has already occurred. It helps us understand how the probability of one event is affected by the occurrence or non−occurrence of another event.

Let's consider two events, A and B. The conditional probability of event A given that event B has occurred is denoted as P(A|B). This is read as "the probability of A given B." The formula for conditional probability is:

$$P(A \mid B) = \frac{P(A \text{ and } B)}{P(B)}$$

Here's a breakdown of the components of the formula:

$P(A \mid B)$: The conditional probability of event A given event B.

$P(A \text{ and } B)$: The probability of both events A and B occurring simultaneously.

$P(B)$: The probability of event B occurring.

111. LEVEL:2

	On time	Late	Total
Route A	5	6	11
Route B	2	10	12
Route C	6	11	17
Total	13	27	40

Victor decides to try three different routes to work for a period of 40 days. In the table above, he tracked whether he arrived to work late or on time each time that he used a particular route. According to the table, what is the probability that Victor took late when he used Route A?

Ⓐ $\frac{3}{20}$　　Ⓑ $\frac{2}{9}$　　Ⓒ $\frac{3}{5}$　　Ⓓ $\frac{6}{11}$

112. LEVEL:3

Language	Native speakers	Non-native speakers	Total
Mandarin Chinese	848	178	1026
Spanish	415	–	–
Hindi	400	200	600
English	335	–	–
Arabic	485	145	625
Total	2483	1237	3570

The table on the left shows the most commonly spoken five languages in the world by native and non-native speakers, in millions, according to the 2013 SIL Ethnologue. If the relative frequency of Spanish non-native speakers to all non-native speakers was 4.8%, approximately how many non-native Spanish speakers, in millions, were there in 2013?

Ⓐ 20　　Ⓑ 61　　Ⓒ 119　　Ⓓ 594

113.

LEVEL:3

	Cheese	No cheese	Total
Sauce	–	–	18
No sauce	–	–	8
Total	20	6	26

Donte and his friends ordered pizza for a birthday party. They asked all the guests at the party whether they wanted sauce or no sauce and whether they wanted cheese or no cheese. The results are displayed in the table above. Donte found that $\frac{1}{5}$ of the people who wanted cheese did not want sauce. What fraction of the people who wanted sauce also wanted cheese?

Ⓐ $\frac{1}{9}$ Ⓑ $\frac{8}{13}$ Ⓒ $\frac{4}{5}$ Ⓓ $\frac{8}{9}$

114.

LEVEL:4

	Fewer than 200 pages	200 pages and above	Total
19th Centry	–	–	–
20th Centry	–	–	14
Total	4	20	–

An English professor classified the novels in his British Literature syllabus by century written and length of the novel. Some of the results are in the table above. If two books are from 19th century and fewer than 200 pages, Based on the table, how many books, both from 20th century that are 200 pages and above, could there be in the class to suggest evidence of association?

3.2 Scatterplots : Linear and Exponential Model

A Scatter Plot has points that show the relationship between two sets of data.
In this example, each dot shows one person's weight versus their height.

Two Quantitative variables ⇒ Bivariate data

1. How to read a scatter plot

The local ice cream shop keeps track of how much ice cream they sell versus the noon temperature on that day. Here are their figures for the last 12 days:

Ice Cream Sales Vs Temperature	
Temperature °C	**Ice Cream Sales**
14.2°	$215
16.4°	$325
11.9°	$185
15.2°	$332
18.5°	$406
22.1°	$522
19.4°	$412
25.1°	$614
23.4°	$544
18.1°	$421
22.6°	$445
17.2°	$408

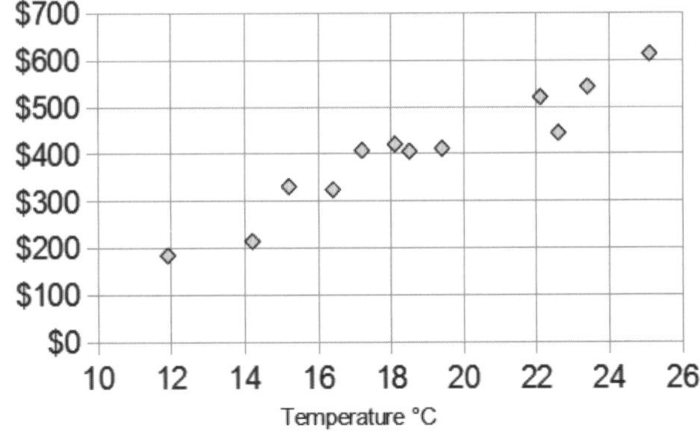

It is now easy to see that warmer weather leads to more sales, but the relationship is not perfect.

2. Patterns of Data in Scatter plots :

⇒ Usual Features : Linearity + Slope + Strength

⇒ Unusual Features : Cluster + Gap + Outlier

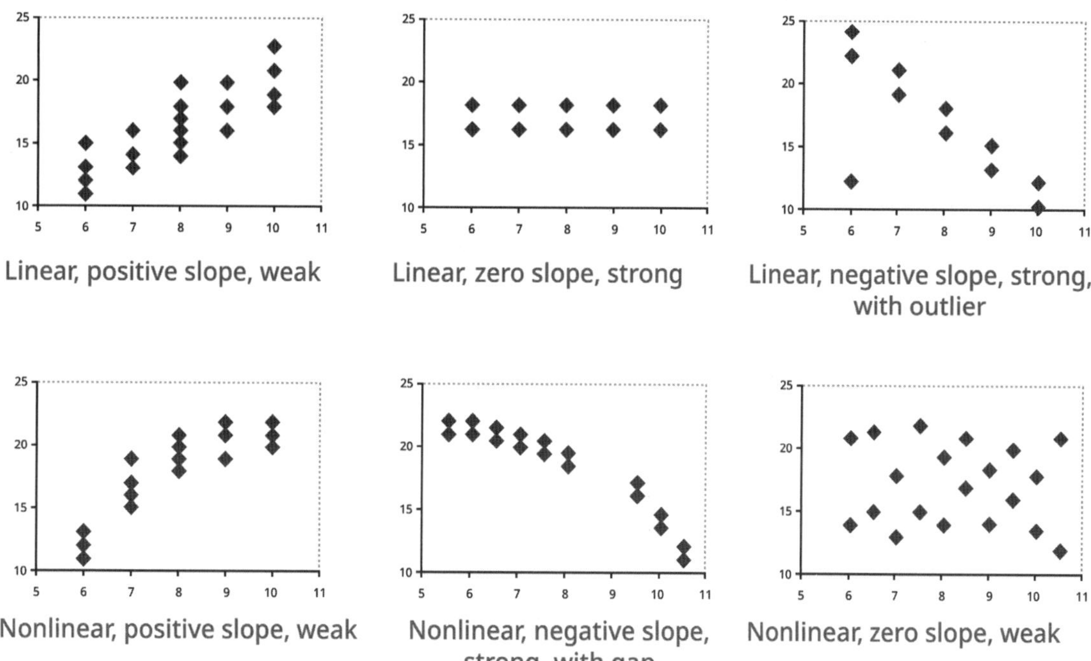

3. Line Regression

Linear regression is used when there is a linear relationship between two variables. It models the relationship using a straight line equation of the form

$$y = mx + b$$

where y represents the dependent variable, x represents the independent variable, m represents the slope of the line, and b represents the y-intercept. In linear regression, the relationship between the variables is assumed to be a constant rate of change.

The goal of linear regression is to find the best-fit line that minimizes the sum of the squared differences between the observed data points and the predicted values on the line. This line can be used to make predictions or estimate values for the dependent variable based on the independent variable.

4. Exponential Regression

Exponential regression is used when there is an exponential relationship between two variables. It models the relationship using an exponential function of the form

$$y = a \times b^x$$

where y represents the dependent variable, x represents the independent variable, a represents the initial value or y-intercept, and b represents the base or growth/decay factor.
Exponential regression is appropriate when the data points exhibit exponential growth or decay patterns. It helps determine the rate of growth or decay and can be used to make predictions or estimate values based on the independent variable.

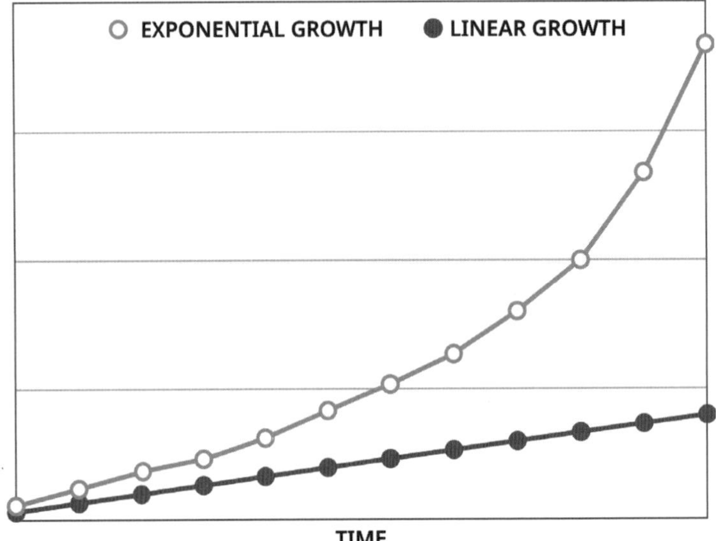

115. LEVEL:2

The scatter plot above shows data for ten charities along with the line of best fit. For the charity with the greatest percent of the total expenses spent on programs when total income is between 3000 and 4000, which of the following is closest to the difference of the actual percent and the percent predicted by the line of best fit?

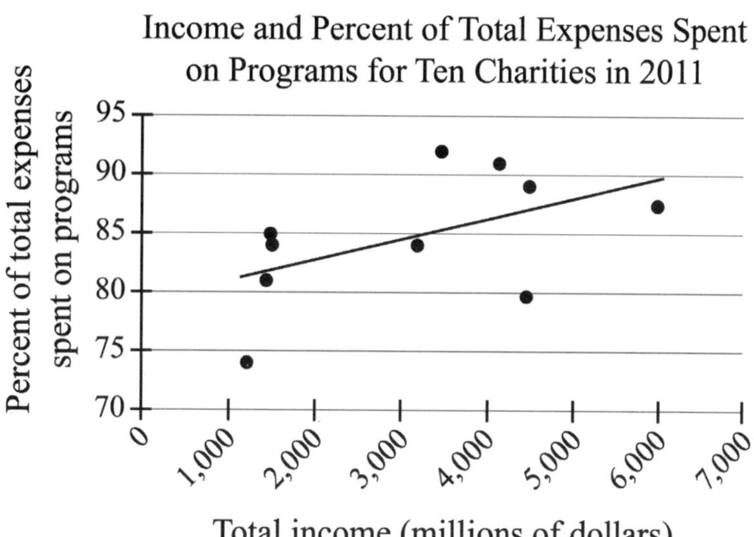

Ⓐ 10%

Ⓑ 7%

Ⓒ 4%

Ⓓ 1%

116. LEVEL:3

U.S. organic food sales

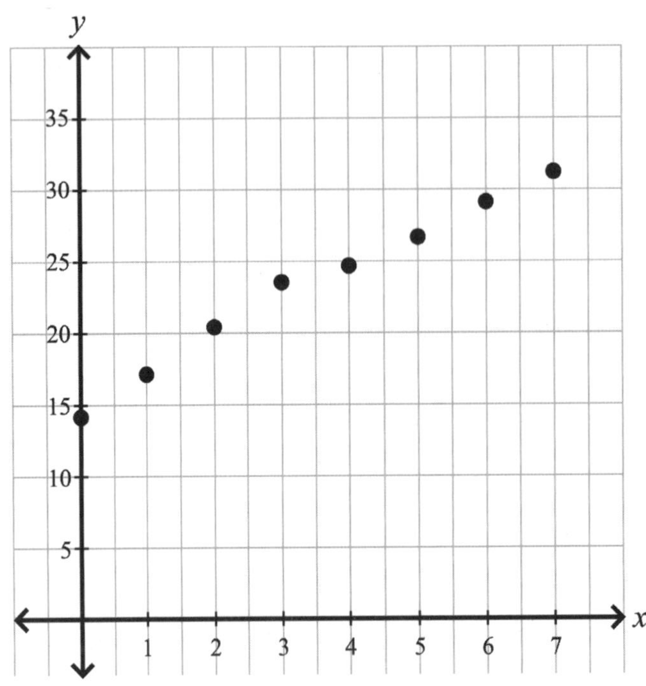

The scatterplot to the left shows information on organic food sales in United States (US) from 2005 to 2012, where x represents years since 2005 and y represents total sales, in billions of dollars, of organic food in the US. Which of the following equations of best models the relationship between the number of years since 2007 and the total sales of organic food?

Ⓐ $y=15.1+0.42x$

Ⓑ $y=15.1+2.4x$

Ⓒ $y=0.42+15.1x$

Ⓓ $y=2.4+15.1x$

117. LEVEL:3

The table is an excerpt of a report released by Statistics Canada in 2008, which show a summary of country-wide agricultural data from the 1986 to 2006.

Census Information from Farms in Canada (1986-2006)					
	1986	1991	1996	2001	2006
Total number of farms	293,089	280,043	276,548	246,923	229,373
Sod					
Area in heetares	20,074	26,797	21,964	22,467	27,960
Total greenhouse products					
Area in square meters	5,176,091	9,306,557	13,437,024	17,567,491	21,697,957

A farmer wants to find out how many farms there will be in 2017 by using a linear function. Which of the following functions correctly approximates the number of the farms in thousands, $f(t)$, in terms of the number of years since 1986, t?

Ⓐ $f(t)=-2.94t+293{,}000$

Ⓑ $f(t)=293t-2.94$

Ⓒ $f(t)=-2.94t+293$

Ⓓ $f(t)=2.94t+293$

118. LEVEL:3

$$P=215(1.005)^{\frac{t}{3}}$$

The equation above can be used to model the population, in thousands, of a certain city t months after 2000. According to the model, the population is predicted to increase by 0.5% every n months. What is the value of n?

119.

LEVEL:3

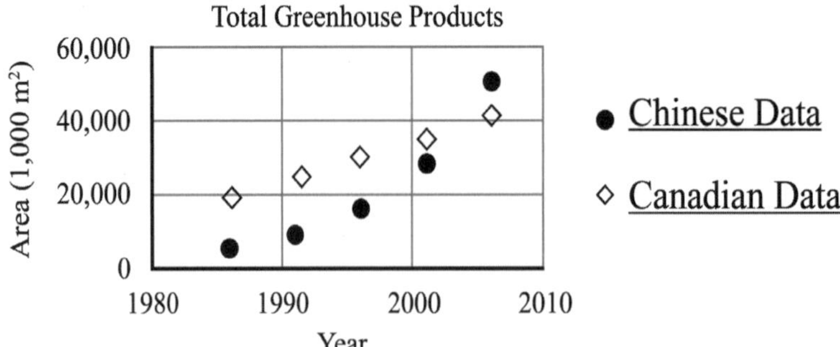

A similar survey is conducted in China, and the results are shown in the scatterplot above. Which of the following staements ia an accurate conclution, based on a comparison between the two countries' greenhouse product growth rates?

Ⓐ Both China's and Canada's production of greenhouse products are growing exponentially.

Ⓑ Both China's and Canada's production of greenhouse products are growing linearly.

Ⓒ China's production of greenhouse products are growing linearly, but and Canada's is growing exponentially.

Ⓓ China's production of greenhouse products are growing exponentially, but and Canada's is growing linearly.

120. LEVEL:3

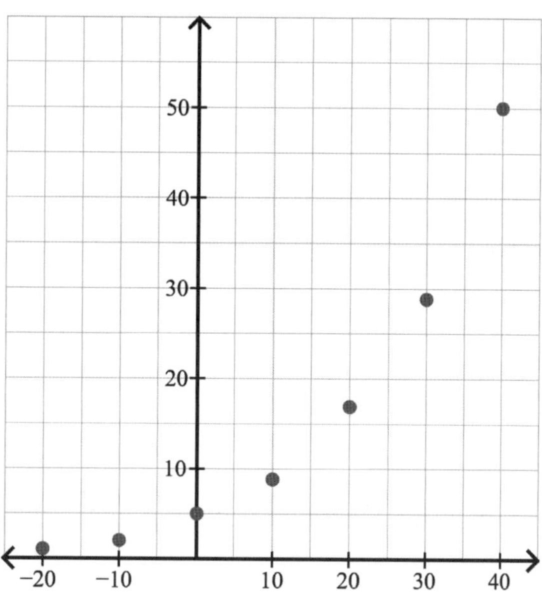

The saturation temperature for water in air is called the dew point. The scatter plot drawn at left shows the amount of water vapor, y, that will saturate 1 cubic meter of air at sea level for different temperatures in degrees Celsius, x. A function that models the data shown is:

$$y = 4.19(1.07)^x$$

What does the value 4.19 in the model tell us about the amount of water vapor that will saturate 1 cubic meter of air at sea level?

Ⓐ It will take approximately 4.19 grams of water vapor to saturate the air when the air temperature is 0 degrees Celsius.

Ⓑ It will take exactly 4.19 grams of water vapor to saturate the air when the air temperature is 0 degrees Celsius.

Ⓒ As the air temperature increases by 1 degree Celsius, the grams of water vapor needed to saturate it increases by a rate of 4.19 %.

Ⓓ As the air temperature increases by 1 degree Celsius, the grams of water vapor needed to saturate it decreases by a rate of 4.19 %.

121. LEVEL:4

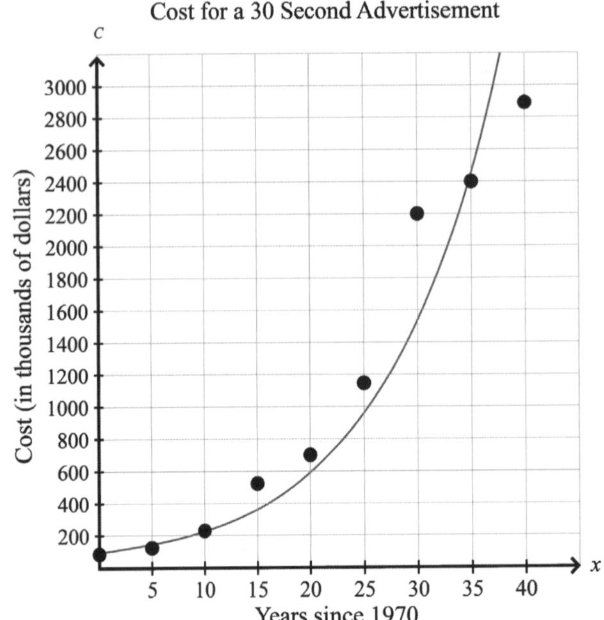

The scatterplot to the left shows the costs to run a 30 second (sec) advertisement during a major sporting event from 1970 to 2010, where x is years since 1970 and c is the cost, in thousands of dollars. A function that models the data shown is:

$$c(x) = 110 \times 1.103^x$$

Based on the model, which of the following is a true statement?

Ⓐ The cost to run a 30 sec advertisement during this sporting event in 1970 was about \$1.1 million.

Ⓑ The cost to run a 30 sec advertisement during this sporting event in 2010 was about \$1.1 million.

Ⓒ Between 1970 and 2010, the cost to run a 30 sec advertisement during this sporting event increased by about \$110,000 each year.

Ⓓ Between 1970 and 2010, the cost to run a 30 sec advertisement during this sporting event increased by about 10% each year.

122. LEVEL:4

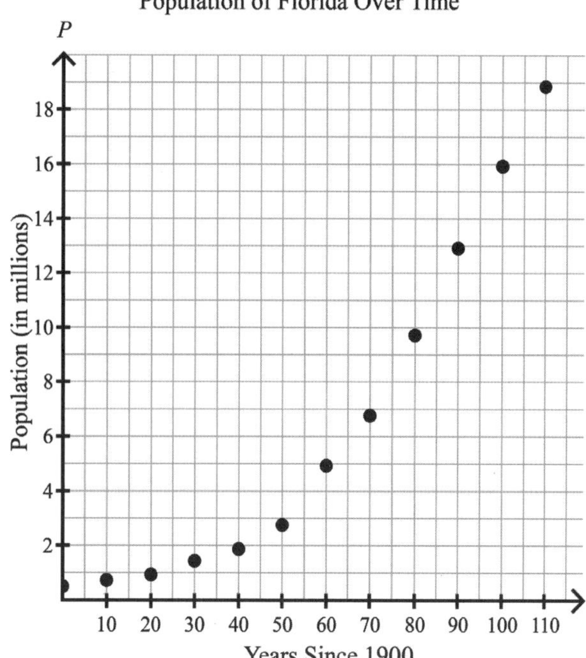

The scatterplot to the left shows the population of Florida, P, in millions, from 1900 to 2010, where t represents years since 1900. Which of the following exponential equations best models the population of Florida from 1900 to 2010?

Ⓐ $P = 0.710\,(0.53)^t$

Ⓑ $P = 0.710\,(0.965)^t$

Ⓒ $P = 0.53\,(1.035)^t$

Ⓓ $P = 0.53\,(1.410)^t$

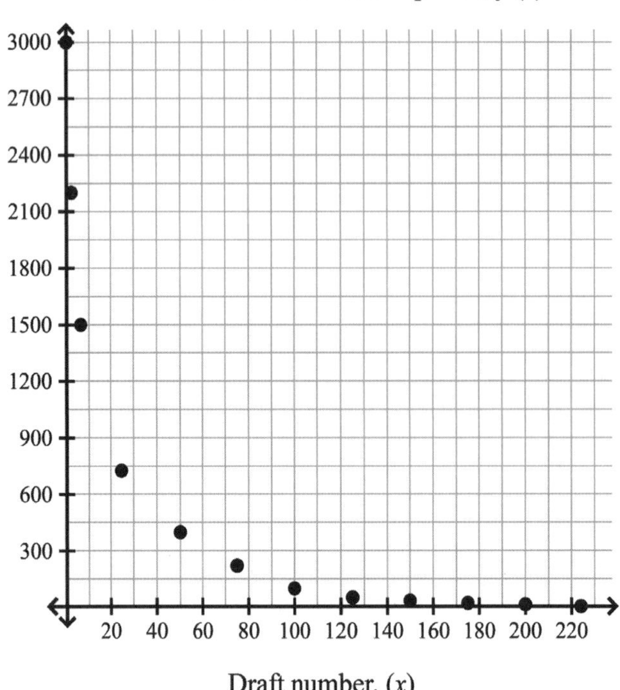

The owners of the teams in the National Football League (NFL) have developed a trade value chart, which assigns a numerical value to recently drafted players. Draftees with the highest numerical value are considered to be most valuable on the trading market. For example, the number 1 draft pick is worth 3,000 draft points and could be traded for picks 2 (worth 2,600 points) and 50 (worth 400 points).

The scatter plot drawn at left depicts the trade value of draftees based on when they are selected in the draft. Which of the following functions best describes the relationship shown?

Ⓐ $f(x) = 0.96 \times (2900.65)^x$

Ⓑ $f(x) = 2900.65 \times (0.96)^x$

Ⓒ $f(x) = 2900.65 \times (9.6)^x$

Ⓓ $f(x) = 960.6 \times (0.29)^x$

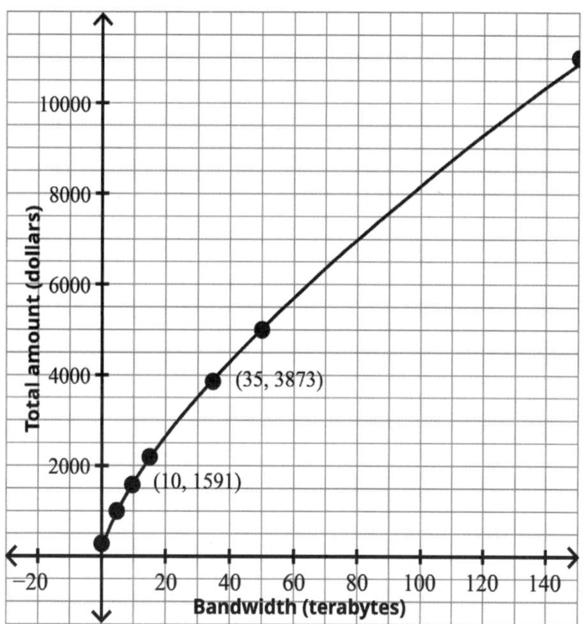

The graph at left in the ta−plane approximates the total amount, a, in dollars, that a particular video hosting company charges a customer who uses t terabytes of bandwidth. What is the best interpretation of the average rate of change between 10 and 35 terabytes?

Ⓐ The company charges about \$91 per terabyte up to the tenth terabyte.

Ⓑ The company charges about \$110 per terabyte up to the tenth terabyte.

Ⓒ The company charges about \$91 per terabyte after the tenth terabyte but before the thirty−fifth.

Ⓓ The company charges about \$110 per terabyte after the tenth terabyte but before the thirty−fifth.

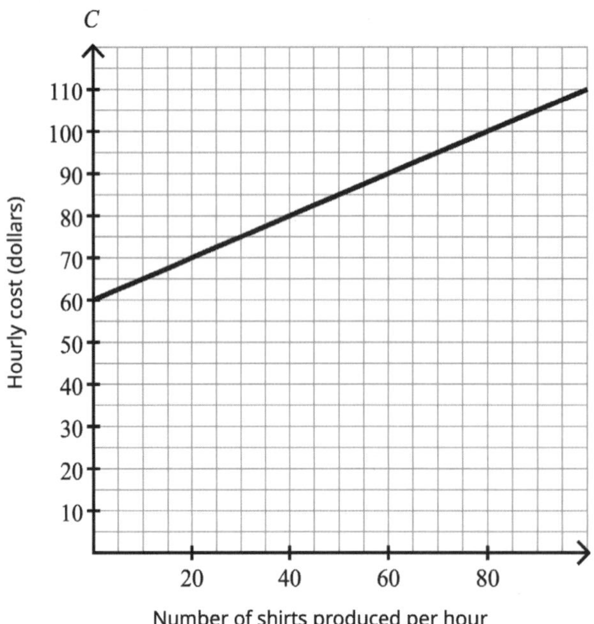

The graph at left in the sC−plane shows the hourly cost C in dollars to a small clothing factory producing s shirts in one hour. Which of the following is correct regarding the average cost per shirt when producings shirts in one hour?

Ⓐ The average cost always increases as the number of shirts produced in one hour increases.
Ⓑ The average cost always decreases as the number of shirts produced in one hour increases.
Ⓒ The average cost reaches a maximum when the number of shirts produced in one hour is 100.
Ⓓ The average cost reaches a minimum of 0.25 dollars per shirt per hour.

126. LEVEL:4

The cumulative costs of a tankless water heater and tank water heater are $4,090 each at the end of the 7th year of use. The cumulative costs of the tankless heater increase $180 per year, but for the tank heater, they increase by 11% per year. By the end of the 10th year, about how much more will the cumulative costs of the tank heater be than of the tankless heater?

Ⓐ $964
Ⓑ $1,504
Ⓒ $4,625
Ⓓ $5,723

127. LEVEL:4

In a particular science research database, the number of indexed genetics articles had increased by an average of 307 articles per year for several years. If there were 6000 genetics articles indexed at the end of that time, and the number of articles increased by 8% annually since then, how many more articles would be indexed than the average annual increase would predict by 2 years later? Round to the nearest article.

128. LEVEL:4

A state accountant models that sales tax revenue grew about 15% per year for several years. Then 2 years ago, when the revenue was $\$2.1$ billion, the revenue began growing by $\$179.5$ million per year instead. About how much less would the revenue be this year than if it had continue d growing 15% per year?

Ⓐ $\$136$ million
Ⓑ $\$318$ million
Ⓒ $\$205$ million
Ⓓ $\$524$ million

129. LEVEL:4

An auto repair company is advertising on television and on the internet. It hires a statistician to poll 5000 people across the country at the beginning of every month for six months. Eight hundred people saw the advertisement on television at the beginning of the first month with an increase of 200 people each month. Also, 1000 people saw the advertisement on the internet at the beginning of the first month with an increase of x percent each month. At the beginning of the third month, there were 240 more people who saw the advertisement on the internet than on television. To the nearest percent, what is the value of x?

130. LEVEL:4

$$\frac{f(t-1)}{f(t)} = 1.0044$$

A mechanical engineer is simulating the rupture of a water tank by puncturing a duplicate water tank and analyzing the water flow. The engineer notices that the rate of water flow, $f(t)$, in liters per second t seconds after puncturing satisfies the equation shown above. Which of the following statements best describes the rate of water flow over time?

Ⓐ The rate of water flow increases linearly.
Ⓑ The rate of water flow decreases linearly.
Ⓒ The rate of water flow increases exponentially.
Ⓓ The rate of water flow decreases exponentially.

131. LEVEL:4

A food safety specialist uses a sensor to detect and measure small electrical currents, in order to determine the concentration of aspartame in soft drinks. The table at left relates the current, in nanoamperes (nA), the sensor detected to the concentration of aspartame, in micromoles per liter. Which of the following statements best describes the relationship?

Current (nA)	Concentration of Aspartame $\left(\frac{\mu\text{mol}}{\text{L}}\right)$
10	0
60	60
110	121
160	182
210	242
260	303
310	363

Ⓐ It is approximately linear because the concentration increases by about the same amount for each $50nA$ increase in current.
Ⓑ It is exactely linear because the concentration increases by about the same percent for each $50nA$ increase in current.
Ⓒ It is approximately exponential because the concentration increases by about the same amount for each $50nA$ increase in current.
Ⓓ It is exactely exponential because the concentration increases by about the same percent for each $50nA$ increase in current.

132. LEVEL:4

A company bought a year of search engine optimization services (SEO) for their site. During that time, the number of visitors per month grew 20% per month, with 500 visitors the final month. After the SEO ended, the site gained 25 visitors per month. How many fewer visitors did the site have 3 months after the SEO ended than there would have been if the number had continued to grow 20% per month?

3.3 Key Features of Graphs

1. Determining the Key Features of Function Graphs

1) What are the key features of graphs?
2) How do I find them?
3) How can I use a graph to tell me information about the function?

2. Key Features of Graphs

1) Rate of Change
2) Domain and Range
3) x-intercepts and y-intercepts
4) Intervals of increasing, decreasing, and constant behavior
5) Parent Equations
6) Maximum value and Minimum value

133. LEVEL:3

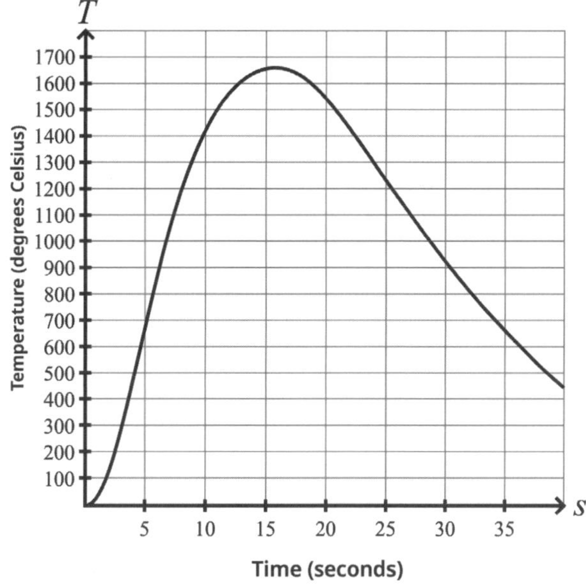

The graph at left in the sT-plane shows the temperature, T, of a *thermite* reaction s seconds after ignition. What is the average rate of increase in temperature per second between $s=0$ and $s=15$ in degrees Celsius per second?

Ⓐ 110

Ⓑ 140

Ⓒ 550

Ⓓ 1650

134.

LEVEL:4

The scatterplot below shows the amount of electric energy generated, in millions of mega watt−hours, by nuclear sources over a 10−year period.

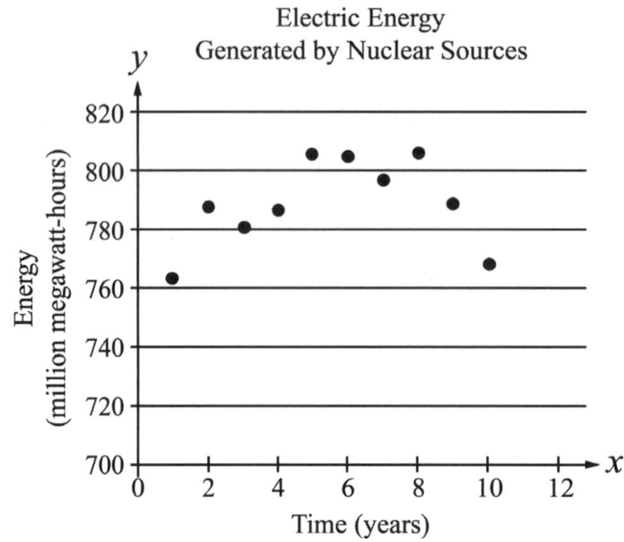

Of the following equations, which best models the data in the scatterplot?

Ⓐ $y = 1.674x^2 + 19.76x - 745.73$

Ⓑ $y = -1.674x^2 - 19.76x - 745.73$

Ⓒ $y = 1.674x^2 + 19.76x + 745.73$

Ⓓ $y = -1.674x^2 + 19.76x + 745.73$

3.4 Center, spread, and shape of distributions

1. Center of Distribution on Data Set :

The center of a distribution refers to the central tendency or the middle point around which the data is clustered. There are several measures used to determine the center:

1) Mean

The mean, or average, is calculated by summing all the values in the data set and dividing by the number of values.

2) Median

The median is the middle value in an ordered data set. If the data set has an even number of values, the median is the average of the two middle values.

Example ▌ Find the median of $3, 7, 10, 20$

$$\text{Median} = \frac{(7+10)}{2} = 8.5$$

3) Mode

The Mode of a list of values is the value or values that appear the greatest number or times. So the mode is the most frequently occurring value in a data set. A distribution can have one mode (unimodal), two modes (bimodal), or more modes (multimodal).

2. Spread of Distribution on Data Set :

The spread, also known as variability or dispersion, measures how the data is distributed or spread out. It provides information about the range or extent of values within the data set. Common measures of spread include:

1) Range: The range is the difference between the maximum and minimum values in the data set.

2) Interquartile range

The interquartile range (IQR) is the distance between the 75th percentile and the 25th percentile. The IQR is essentially the range of the middle 50% of the data. Because it uses the middle 50%, the IQR is not affected by outliers or extreme values.

3) Standard Deviation: The standard deviation measures the average distance between each data point and the mean. It provides a measure of how spread out the data is from the mean.

135. LEVEL:2

A data set consists of the values 250, 320, 430, 452, 600 and $17,167$. If the outlier is removed, what will happen to the value of the mean of the data set?

Ⓐ The mean will remain the same.

Ⓑ The mean will decrease.

Ⓒ The mean will increase.

Ⓓ There in not enough information to determine how the mean will change.

136. LEVEL:3

For a certain computer game, individuals receive an integer score that ranges from 2 through 10. The integer score that ranges from 2 through 10. The table below shows the frequency distribution of the scores of the 9 players in the group A and the 11 players in the group B.

	Score	2	3	4	5	6	7	8	9	10	Total
Score equencies	Group A	1	1	2	1	3	0	0	1	0	9
	Group B	0	0	0	4	2	0	2	1	2	11

The mean of the score for group A is 5. and the mean of the score of group B is 7. What is the mean of the scores for the 20 players in group A and B combined?

137. LEVEL:3

Ticket Prices by Row Number

Row number	Ticket price
1–2	\$25
3–10	\$20
11–20	\$15

The price of a ticket to a play is based the row the seat is in, as shown in the table above. A group wants to purchase 10 tickets for the play.

They will purchase 3 tickets for seats in the row 1.

They will purchase 2 tickets for seats in the row 3.

They will purchase 2 tickets for seats in the row 4.

They will purchase 3 tickets for seats in the row 12.

What is the average (arithmetic mean) ticket price, in dollars, for the 10 tickets? (Disregard the \$ sign when gridding your answer.)

138. LEVEL:3

A fish hatchery has three tanks for holding fish before they are introduced into the wild. Ten fish weighting less than 5 ounces are placed in tank A. Eleven fish weighing at least 5 ounces but no more than 13 ounces are placed in tank B. Twelve fish weighing more than 13 ounces are placed in tank C. Which of the following could be the median of the weights, in ounces, of these 33 fish?

Ⓐ 4.5

Ⓑ 8

Ⓒ 13.5

Ⓓ 15

139. LEVEL:4

Land surveyors visited a small fishing village and divided the land into plots, each 120 square meters in area. They counted the number of dwellings on each plot and recorded their data in the bar graph to the left. Due to people moving away from the village, some residents are now combining dwellings together in order to create larger ones. If this is the only change being made to the data, then which of the following must be true when the land is surveyed again?

Ⓐ The mean number of dwellings will decrease.
Ⓑ The median number of dwellings will decrease.
Ⓒ The range of the number of dwellings will increase.
Ⓓ The variance of the number of dwellings will increase.

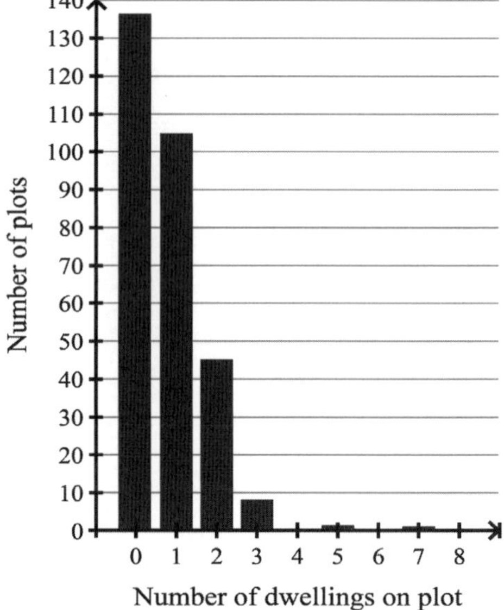

140. LEVEL:4

Judah, a marketing consultant, tracked how many times his customers asked a **question** during his sales calls over a one month period. Based on the results shown in the distribution above, what was the median number of questions asked?

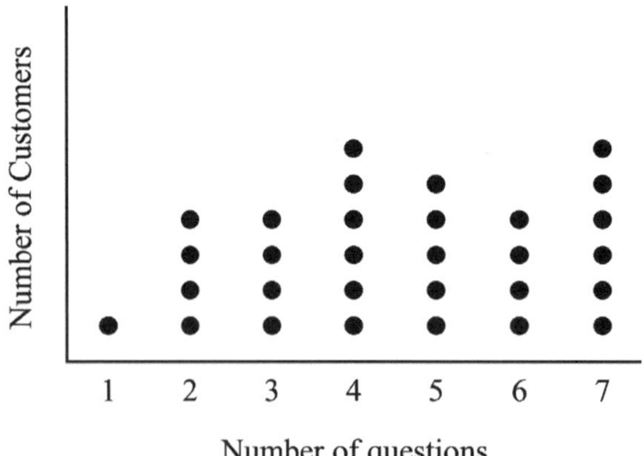

Ⓐ 4 Ⓑ 4.5 Ⓒ 5 Ⓓ 5.5

141.

LEVEL:4

The 22 students in a health class conducted an experiment in which they recorded their pulse rates, in beats per minute, before and after completing a light exercise routine. The plots below display the results..

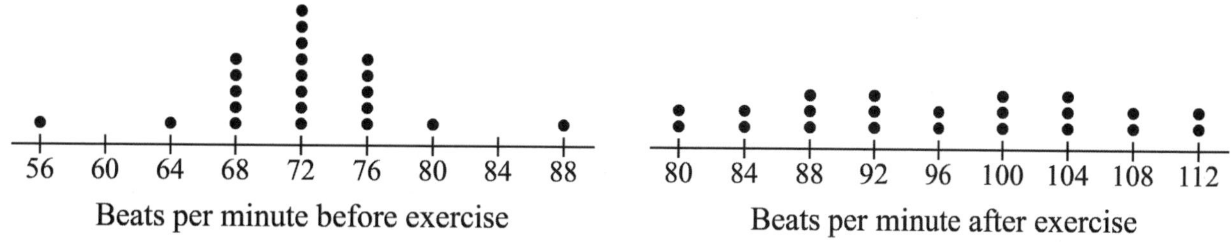

Let s_1 and r_1 be the standard deviation and range, respectively, of the data before exercise, and let s_2 and r_2 be the standard deviation and range, respectively, of the data after exercise. Which of the following is true?

Ⓐ $s_1 = s_2$ and $r_1 = r_2$

Ⓑ $s_1 < s_2$ and $r_1 < r_2$

Ⓒ $s_1 > s_2$ and $r_1 > r_2$

Ⓓ $s_1 \neq s_2$ and $r_1 = r_2$

142.

LEVEL:4

A class used catapults to launch two kinds of gummy candies. The dot plots at left record the distances the gummy candies travelled, in inches. Which statement best compares the standard deviations and the means of the distances travelled of the two kinds of candies?

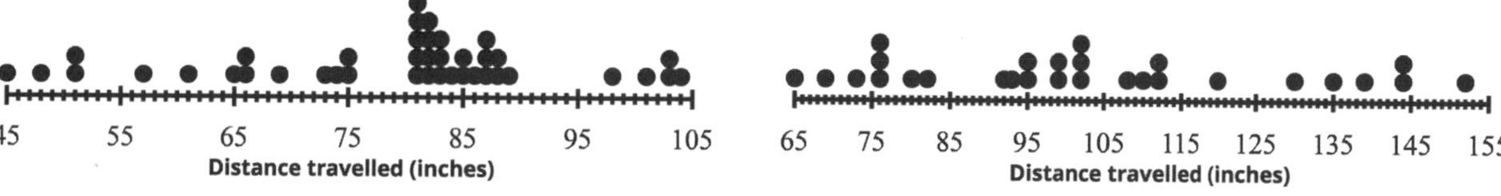

Ⓐ The worm gummy distances have a greater standard deviation and mean than the fish gummy distances.

Ⓑ The worm gummy distances have a greater standard deviation, but a lower mean than the fish gummy distances.

Ⓒ The fish gummy distances have a greater standard deviation and mean than the worm gummy distances.

Ⓓ The fish gummy distances have a greater standard deviation, but a lower mean than the worm gummy distances.

143. LEVEL:4

An employer wanted to compare the commute times between 1st and 2nd shift employees. Both shifts had a mean commute of 17 minutes. The histograms to the left summarize the average commute times of the employees. Which of the following statements best compares the standard deviations of the shifts?

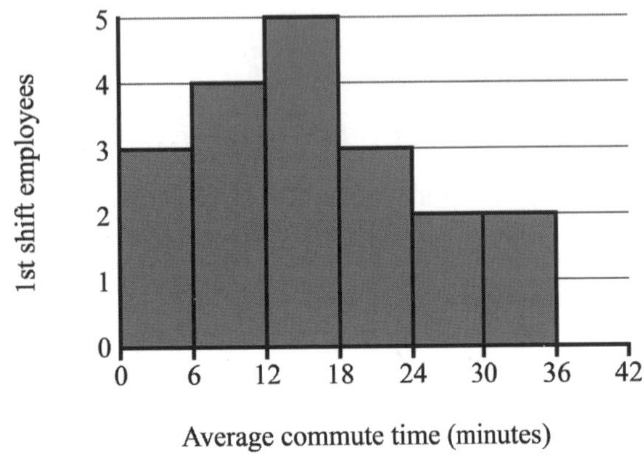

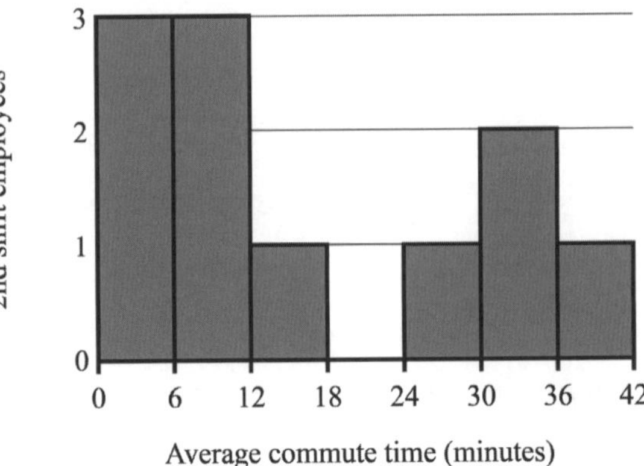

Ⓐ The standard deviations are equal.

Ⓑ The standard deviation for the 1st shift employees is greater.

Ⓒ The standard deviation for the 2nd shift employees is greater.

Ⓓ Histograms do not provide enough information to compare standard deviations.

144. LEVEL:4

The histograms to the left show the test results of two different 11th grade history classes on the same exam. Which statement best compares the standard deviations of the two data sets?

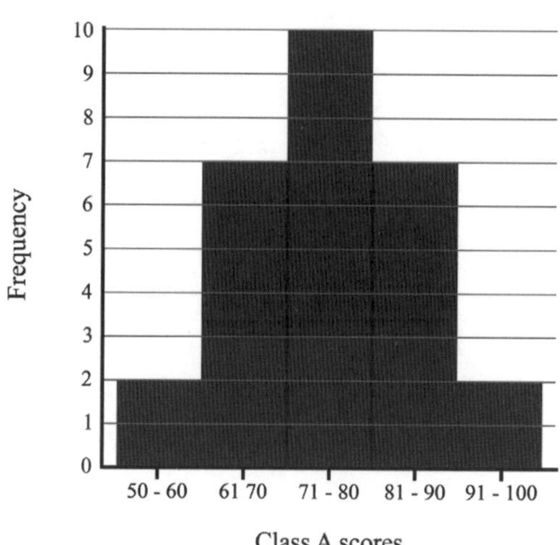

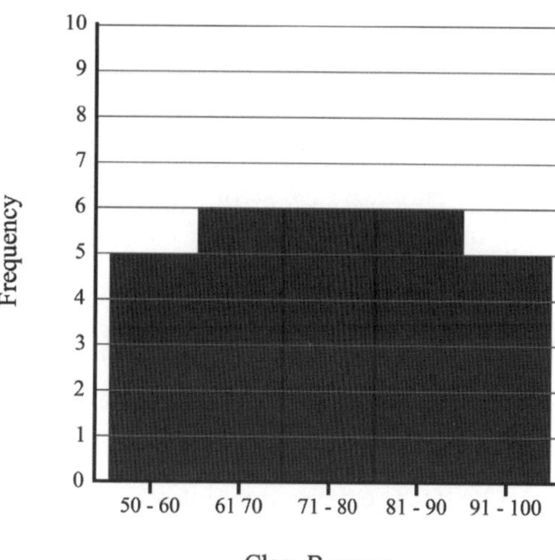

Ⓐ The standard deviation of the scores from class A is greater than the standard deviation of the scores from class B.

Ⓑ The standard deviation of the scores from class B is greater than the standard deviation of the scores from class A.

Ⓒ The standard deviation of the scores from class A is equal to the standard deviation of the scores from class B.

Ⓓ There is not enough information to compare the two standard deviations.

3.5 Data collection and conclusions

1. Margin of Error

The margin of error is a statistical concept used to quantify the amount of uncertainty or potential error associated with survey or sample data. It provides a range within which the true population value is likely to fall. Understanding the margin of error is important for interpreting and making conclusions based on survey results.

The margin of error is typically represented as a range or interval around a sample statistic, such as a mean or proportion. It is calculated based on the sample size, variability of the data, and desired level of confidence. The margin of error provides a measure of how close the sample statistic is expected to be to the true population value.

Example ▮

Suppose a survey is conducted to estimate the proportion of adults who support a particular political candidate. A random sample of 500 adults is taken, and the survey finds that 55% of the respondents support the candidate. The margin of error for the survey may be calculated as $\pm 3\%$, with a 95% confidence level. This means that we can be 95% confident that the true proportion of adults who support the candidate falls within the range of 52% to 58% $(55\% \pm 3\%)$.

Margin of Error (Parameter) = Critical Vaule × Standard deviation for the population.

Margin of Error (Statistic) = Critical Vaule × Standard error for the sample.

2. Data collection and conclusions

Data collection and drawing conclusions are important components of data analysis in SAT Math. These skills involve understanding how data is collected, evaluating the reliability and validity of data sources, and making logical inferences based on the data.

1) Data Collection:

Data collection involves gathering information or observations to analyze and draw conclusions. Data can be collected through various methods, such as surveys, experiments, observations, or existing sources. When evaluating data collection methods, it's important to consider factors like sample size, sampling methods, bias, and the representativeness of the data.

2) Data Sources:

On the SAT Math, you may encounter questions that provide data sets or graphs representing collected data. It's important to evaluate the reliability and validity of the data sources. Consider factors such as the source of the data, potential biases, and the methodology used to collect the data. Understanding the context and limitations of the data helps you make accurate conclusions.

3) Drawing Conclusions:

Drawing conclusions from data involves making logical inferences based on the information provided. This may involve analyzing patterns, trends, or relationships between variables. Statistical techniques such as calculating means, medians, or correlation coefficients can be used to support conclusions.

4) Causation vs. Correlation:

It's important to differentiate between causation and correlation when drawing conclusions. Correlation indicates a relationship between variables, but it does not imply causation. It's crucial to be cautious when making causal claims based on correlation alone. Consider other evidence or experimental designs to establish causation.

5) Generalization:

When drawing conclusions from data, consider the extent to which the conclusions can be generalized to a larger population or context. If the data represents a specific sample, be cautious about applying the conclusions to a broader population without proper justification.

145. LEVEL:3

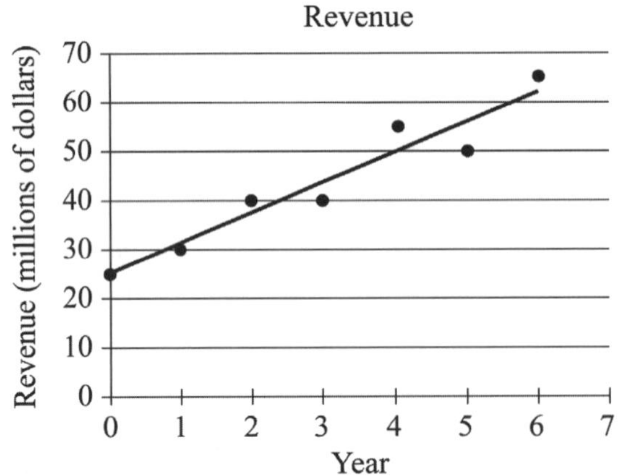

The scatterplot above shows the revenue, in millions of dollars, that a company earned over several years and a line of best fit for the data. In year 4, the difference between the actual revenue and the predicted revenue is n million dollars, where n is a positive integer. What is the value of n? Round your answer to the nearest whole number. (Disregard the \$ sigh when griding your answer.)

146. LEVEL:4

Sample	Percent in favor	Margin of error
A	52%	4.2%
B	48%	1.6%

The results of two random samples of votes for a proposition are shown above. The samples were selected from the same population, and the margins of error were calculated using the same method. Which of the following is the most appropriate reason that the margin of error for sample A is greater than the margin of error for sample B?

Ⓐ Sample A had a smaller number of votes that could not be recorded.

Ⓑ Sample A had a higher percent of favorable reponses.

Ⓒ Sample A had a larger sample size.

Ⓓ Sample A had a smaller sample size.

147.

LEVEL:4

An ecologist selected a random sample of 30 prairie dogs from a colony and found that the mean mass of the prairie dogs in the sample was 0.94 kilograms (kg) with an associated margin of error 0.12 kg. Which of the following os the best interpretation of the ecologist's finding?

Ⓐ All prairie dogs in the sample have a mass between 0.82 kg and 1.06 kg.

Ⓑ Most prairie dogs in the sample have a mass between 0.82 kg and 1.06 kg.

Ⓒ Any mass between 0.82 kg and 1.06 kg is a plausible value for the mean mass of the prairie dogs in the sample.

Ⓓ Any mass between 0.82 kg and 1.06 kg is a plausible value for the mean mass of the prairie dogs in the colony.

148.

LEVEL:4

An ecologist conducted measured the population of brown bears in a North American region and the number of deforested acres in the same region since the year 2000.

The study concluded that as the population of brown bears steadily decreased, the number of deforested acres steadily increased during the same time period. Based on this data, which conclusion is valid?

Ⓐ The increase in the number of deforested acres in the North American region since 2000 caused the decrease in the brown bear population there during the same time period.

Ⓑ The decrease in the brown bear population in the North American region since 2000 caused the increase in the number of deforested acres there during the same time period.

Ⓒ There is no evidence of an association between the brown bear population levels in the North American region and the number of deforested acres there in the years since 2000.

Ⓓ There is evidence of an association between the brown bear population levels in the North American region and the number of deforested acres there in the years since 2000.

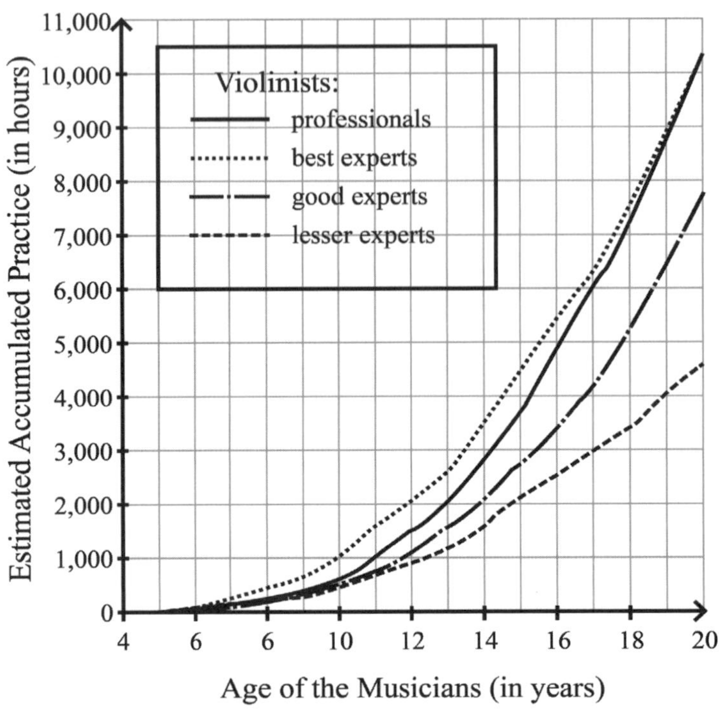

Adapted from "The Role of Deliberate Practice in the Acquisition of Expert Performance," by K.A. Ericsson, R. Th. Krampe, and C. Tesch-Romer, 1993, Psychological Review, 700(3).

In a famous study on the role of practice in the acquisition of expert performance, psychologists compared the amount of time spent on solitary practice, based on diaries and retrospective estimates, for four groups of violinists: professional violinists, the best expert violinists, good expert violinists, and the least accomplished expert violinists (lesser experts). Based on the results of this study, which conclusion is best supported by the data?

Ⓐ A violinist who practices about 10,000 hours by the age of 20 will become a professional violinist.

Ⓑ By the age of 20, the best experts and professional violinists in the study had practiced more than twice as much as the least accomplished violinists.

Ⓒ The least accomplished violinists did not practice as much because they became discouraged.

Ⓓ There is no evidence of an association between increased solitary practice before the age of 18 and level of expertise as a violinist.

150.
LEVEL:4

A local tv news station wants to determine how often and through which medium their viewers check the weather. Which of the following survey methods is most likely to produce valid results?

Ⓐ Ask a random sample of their viewers how much they enjoy the weather portion of the local news.
Ⓑ Ask a random sample of their viewers whether they own a smartphone.
Ⓒ Ask a random sample of members of the local meteorological society whether they watch the local news.
Ⓓ Ask a random sample of their viewers how often and when they use various sources to obtain weather information.

151.
LEVEL:4

A writer for a high school newspaper is conducting a survey to estimate the number of students that will vote for a particular candidate in an upcoming student government election. All students at the high school are eligible to vote in the election, and the writer decides to select a sample of students to take the survey. Which of the following sampling methods is most likely to produce valid results?

Ⓐ Survey every fifth student to enter the school library.
Ⓑ Survey every fifth student to arrive at school one morning.
Ⓒ Survey every fifth senior to arrive at school one morning.
Ⓓ Survey every fifth student to enter the school stadium for a football game.

152. LEVEL:4

For the finale of a TV show, viewers could use either social media or a test message to vote for their favorite or two contestants. The contestant receiving more than 50% of the vote won. An estimated 10% of the viewers voted, and 30% of the votes were cast on social media. Contestant 2 earned 70% of the votes cast using social media and 40% of the votes cast using a test message. Based on this information, which of the following is an accurate conclusion?

Ⓐ If all viewers had voted, Contestant 2 would have won.

Ⓑ Viewers voting by social media were likely to be younger than viewers voting by text message.

Ⓒ If all viewers who had voted had voted by social media instead of by text message, Contestant 2 would have won.

Ⓓ Viewers voting by social media were likely to prefer Contestant 2 than more viewer voting by text message.

Additional Topics in Math:

Geometry, Coordinate geometry and Trigonometry

Circle to square and cubes to double
would give a man excessive trouble.
-MATTHEW PRIOR (1664-1721)

☑ Geometry, Coordinate geometry and Trigonometry section of the SAT Math

The Geometry, Coordinate Geometry, and Trigonometry sections on the SAT Math cover different aspects of these mathematical topics. Here's an explanation of each section:

1. Geometry :

The Geometry section focuses on the properties, relationships, and measurements of shapes and figures. It includes concepts such as angles, lines, polygons, circles, triangles, quadrilaterals, and three-dimensional figures. You may be asked to solve problems involving area, perimeter, volume, surface area, congruence, similarity, and transformations. It is important to understand geometric formulas, theorems (e.g., Pythagorean Theorem), and postulates, as well as how to apply them to solve problems.

2. Coordinate Geometry :

Coordinate Geometry involves using coordinate planes and algebraic equations to describe and analyze geometric figures. You'll work with points, lines, slopes, distances, midpoints, parallel and perpendicular lines, and equations of lines. You may be asked to determine the equation of a line, find the coordinates of a point, calculate distances between points, or analyze geometric properties using coordinate geometry techniques.

3. Trigonometry :

Trigonometry deals with the relationships between angles and sides of triangles. In this section, you'll work with trigonometric functions such as sine, cosine, and tangent, as well as inverse trigonometric functions. You'll apply trigonometry to solve problems involving right triangles, including finding missing side lengths and angles. It's important to understand trigonometric ratios, the unit circle, special triangles (e.g., 30−60−90 and 45−45−90 triangles), and how to use trigonometric functions to solve problems.

To excel in these sections, it's essential to review and practice a variety of geometric concepts, including properties of shapes, formulas, and theorems.

4.1 Volume word problems including lengths and areas

1. Volume of a Cube

$$V = a^3$$

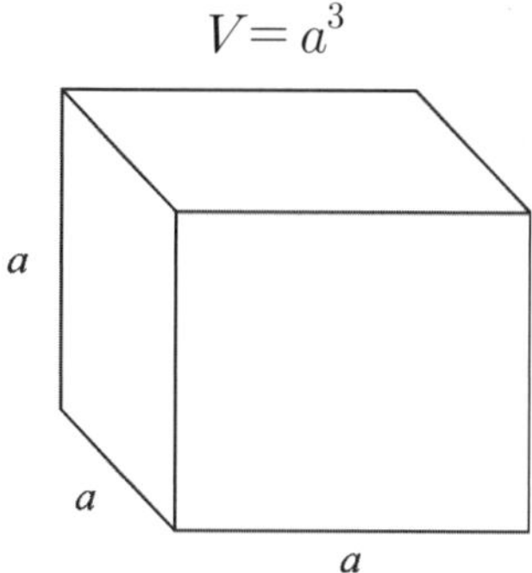

(a is the length of the side of each edge of the cube)

2. Volume of a Rectangular prism

$$V = abc$$

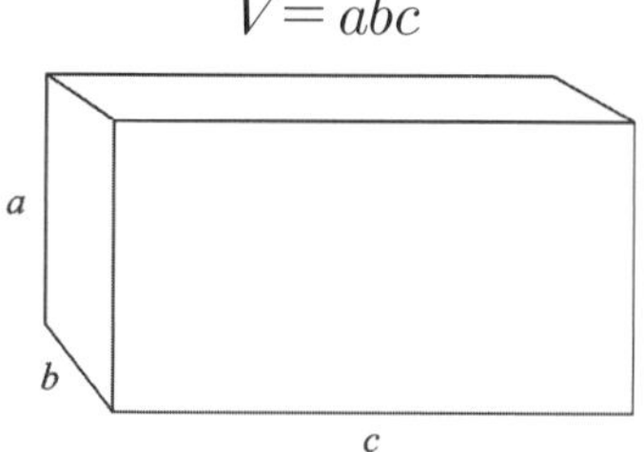

(a, b, and c are the lengths of the 3 sides)

3. Volume of a Irregular prism

$$V = bh$$

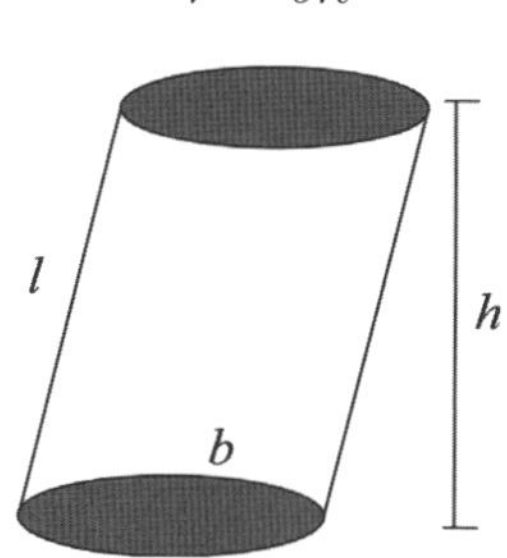

(b: Surface of base , h: height)

4. Volume of a Cylinder

$$V = bh = \pi r^2 h$$

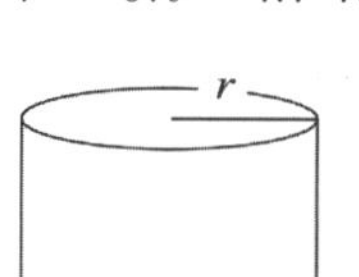

Cylinder

(b: Surface of base , h: height)

5. Volume of a Pyramid

$$V = \frac{1}{3}bh$$

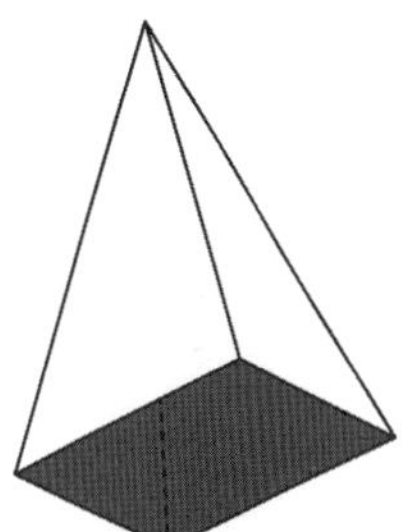

Pyramid

(b: Surface of base , h: height)

6. Volume of a Cone

$$V = \frac{1}{3}\pi r^2 h$$

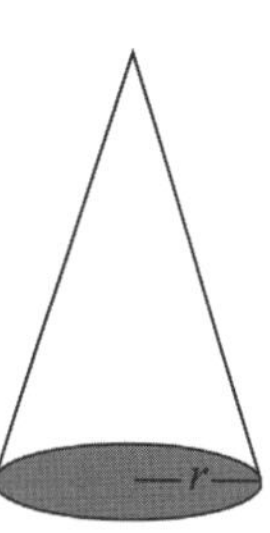

Cone

7. Volume of a Sphere

$$V = \frac{4}{3}\pi r^3$$

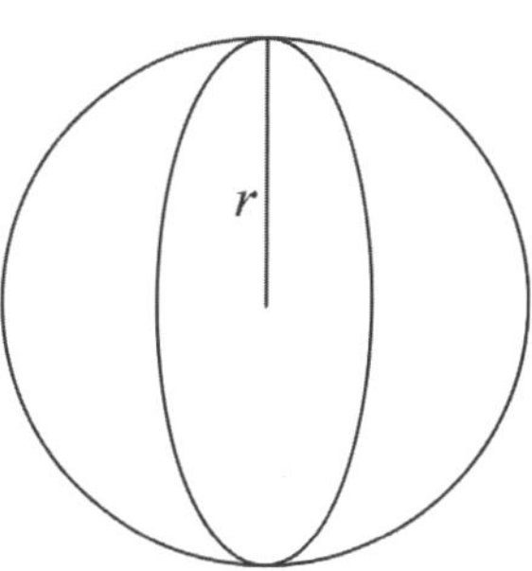

Sphere

(r is radius of circle)

153. LEVEL:4

In triangle ABC, the measure of angle A is $50°$. If triangle ABC is isosceles, which of the following is NOT a possible measure of angle B?

Ⓐ $50°$

Ⓑ $65°$

Ⓒ $80°$

Ⓓ $100°$

154. LEVEL:4

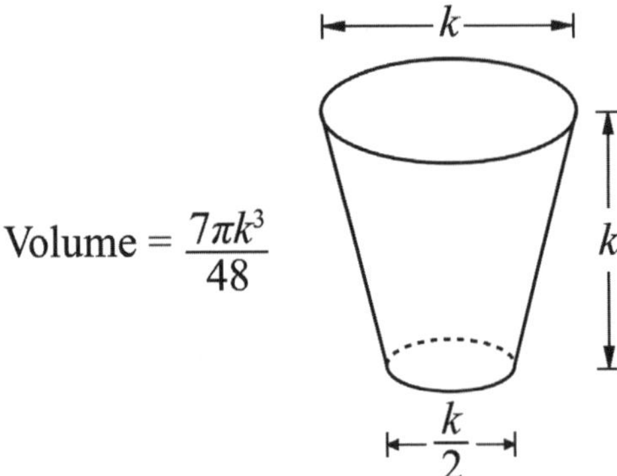

The glass pictured above can hold a maximum volume of 473 cubic centimeters, which is approximately 16 fluid ounces. what is the value of k, in centimeters?

Ⓐ 2.52

Ⓑ 7.67

Ⓒ 7.79

Ⓓ 10.11

155. LEVEL:4

A pyramid is formed from six identical steel beams of length l such that its base forms an equilateral triangle. The height is equal to h. What is the height of the pyramid in terms of l?

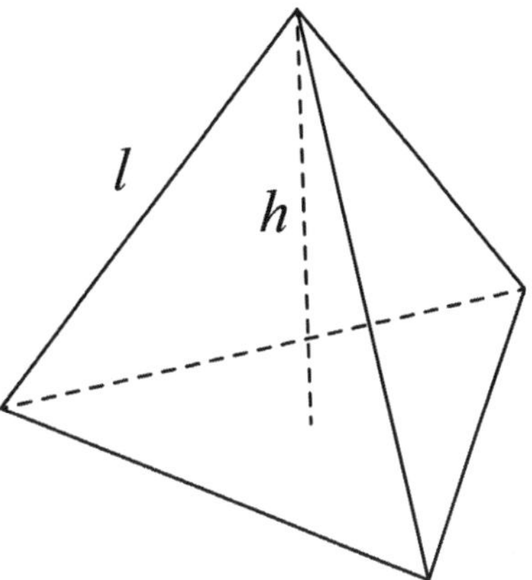

Ⓐ $\frac{\sqrt{2}}{12}l$ Ⓑ $\frac{\sqrt{3}}{12}l$ Ⓒ $\frac{\sqrt{2}}{4}l$ Ⓓ $\frac{\sqrt{6}}{3}l$

156. LEVEL:4

A pyramid is formed from six identical steel beams of length l such that its base forms an equilateral triangle. The height is equal to h. What is the volume of the pyramid in terms of l?

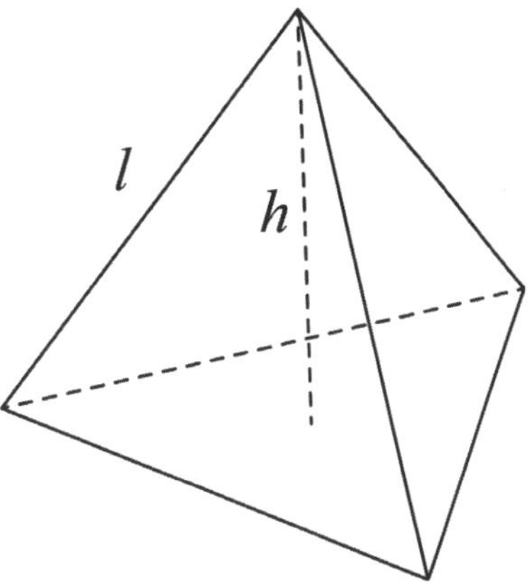

Ⓐ $\frac{\sqrt{2}}{12}l^3$ Ⓑ $\frac{\sqrt{3}}{12}l^3$ Ⓒ $\frac{\sqrt{2}}{4}l^3$ Ⓓ $\frac{\sqrt{3}}{4}l^3$

157. LEVEL:4

An oxygen tank is shaped like a cylinder with two hemispheres attached to the cylinder bases as shown at left. The diameter of the cylinder and hemispheres is 6 inches(in). The height of the entire tank is16in. What is the volume of the oxygen tank in cubic inches?

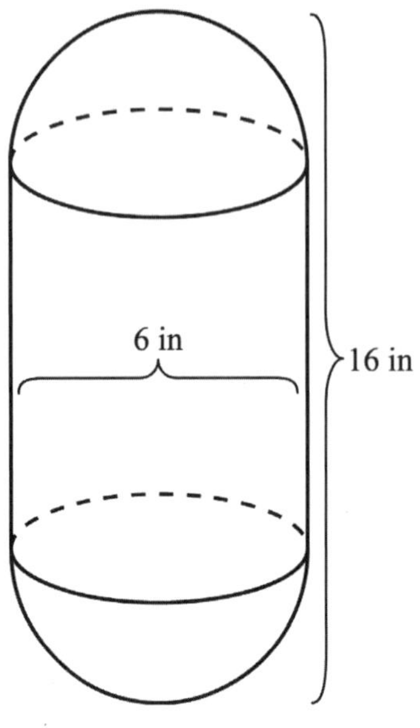

Ⓐ 126π Ⓑ 144π Ⓒ 180π Ⓓ 504π

158. LEVEL:4

A metal pipe has the shape of a hollow cylinder as shown at left. The length of the pipe is $l=30$ centimeters (cm), the thickness of the pipe is $t=1cm$, and the internal diameter is $d=4cm$. What is the volume of the pipe material in cubic centimeters?

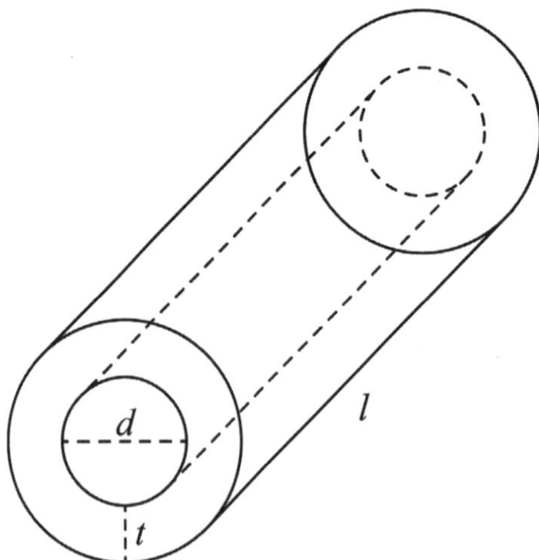

Ⓐ 120π Ⓑ 150π Ⓒ 270π Ⓓ 600π

159. LEVEL:4

A regulation FIFA soccer ball is modeled after the Buckyball, that was introduced by Buckminster Fuller. Due to air pressure, the inflated ball assumes the shape of an almost perfect sphere. Aregulation soccer ball has a circumference of 70 centimeters(cm). What is the approximate volume of the soccer ball in cubic centimeters(cm^3)? Use $\pi \approx 3.14$ and round the answer to the nearest whole number.

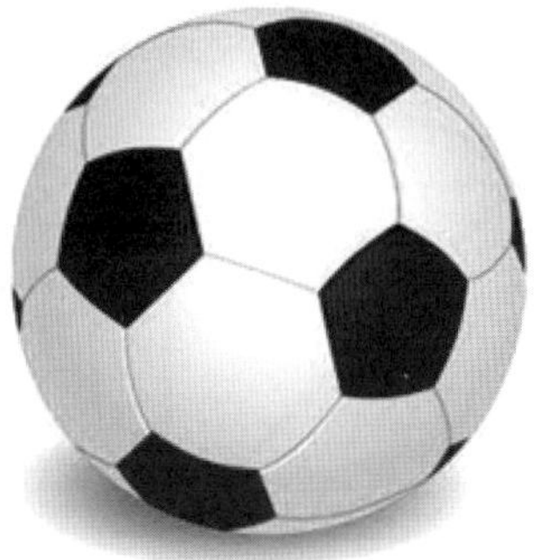

Ⓐ 5,792 cm^3 Ⓑ 46,338 cm^3 Ⓒ 179,594 cm^3 Ⓓ 1,436,755 cm^3

160. LEVEL:4

A "conical bottom tank," shown at the bottom, is a large container used in water purification plants. The tank consists of a cylinder with a cone at the bottom for drainage. A conical bottom tank has a cylinder of height 175 centimeters and diameter 120 centimeters. The cone at the bottom has height 50 centimeters. What is the volume of the conical bottom tank? Use $\pi \approx 3.14$.

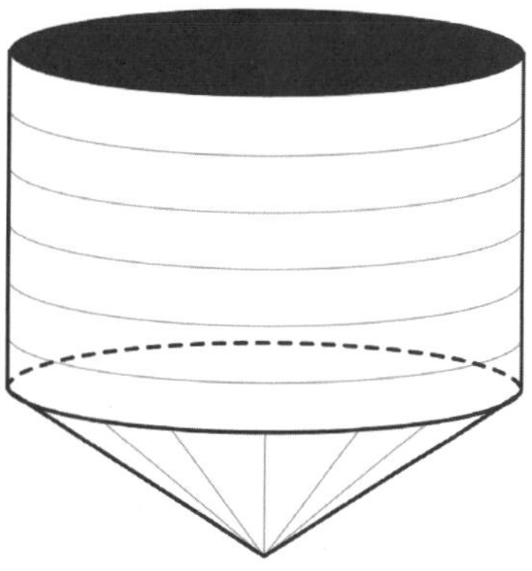

Ⓐ 1,789,800 cubic centimeters
Ⓑ 188,400 cubic centimeters
Ⓒ 1,978,200 cubic centimeters
Ⓓ 2,167,700 cubic centimeters

161.

LEVEL:4

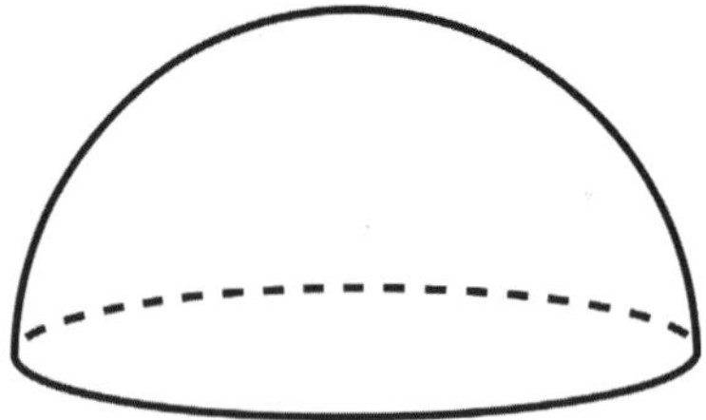

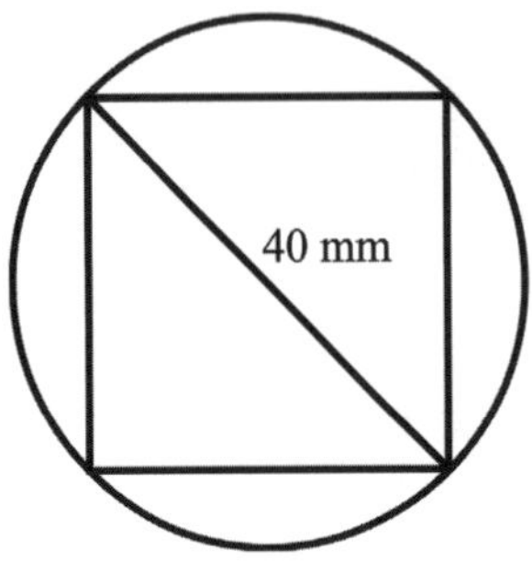

A gemstone, shown at left, is being carved in the shape of a right square pyramid from a hemispherical stone with diameter 40 millimeters(mm). The square base of the pyramid will be carved from the base of the hemisphere. The height of the pyramid will coincide with the radius of the hemisphere. What is the volume, in cubic millimeters(mm^3), of the right square pyramid that can be carved from the hemisphere, as shown below?

Ⓐ 2,667 mm^3 Ⓑ 5,333 mm^3 Ⓒ 10,667 mm^3 Ⓓ 16,746 mm^3

4.2 Right triangle word problems

1. Pytagora's Theorem (Pythagorean Theorem)

In a right triangle, the square of the length of the hypotenuse (the side opposite the right angle) is equal to the sum of the squares of the lengths of the other two sides. This can be written as:

$$a^2 + b^2 = c^2$$

where a and b are the lengths of the two shorter sides (legs) of the right triangle.
c is the length of the hypotenuse.

2. The Distance Between Two Points

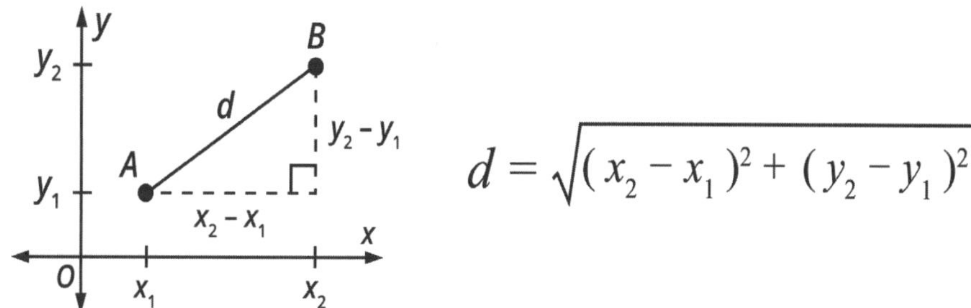

As discussed, the distance formula is used to find the distance between any two points, when we already know the coordinates. The points could be present alone on the x-axis or y-axis or in both axes.

Let us consider, there are two points say A and B in an XY plane. The coordinates of point A are (x_1, y_1) and of B are (x_2, y_2).

$$d = \sqrt{(x_2 - x_1)^2 + (y_2 - y_1)^2}$$

Note: If the coordinates of two points P and Q are such that, $(x_1, 0)$ and $(x_2, 0)$, the distance between PQ will be given by:

$$PQ = |x_2 - x_1|$$

3. Special Right Triangles

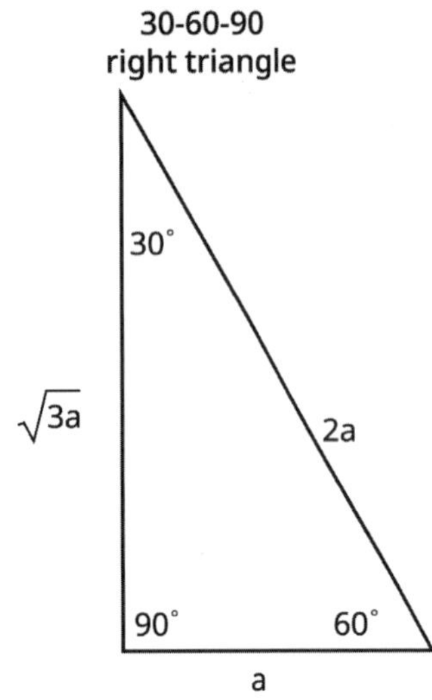

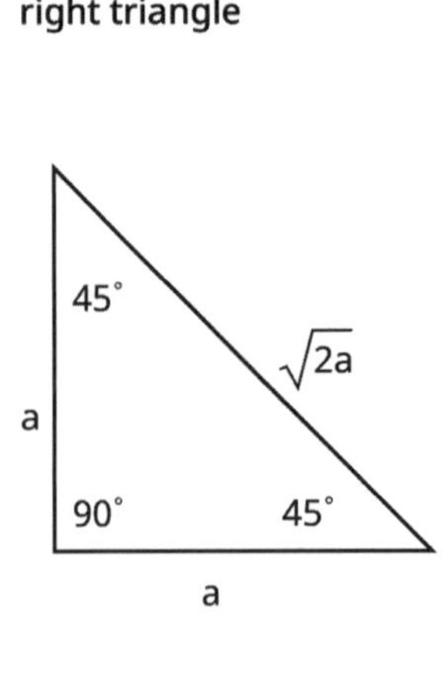

1) $30-60-90$ Triangle:

A $30-60-90$ triangle is a right triangle with one angle measuring 90 degrees, one angle measuring 30 degrees, and the third angle measuring 60 degrees. The ratio of the side lengths in this triangle is:

Shorter leg length : Longer leg length : Hypotenuse length

$$1:\sqrt{3}:2$$

2) $45-45-90$ Triangle:

A $45-45-90$ triangle, also known as an isosceles right triangle, has two congruent (equal) acute angles of 45 degrees each. The ratio of the side lengths in this triangle is:

Leg length : Leg length : Hypotenuse length

$$1:1:\sqrt{2}$$

162. LEVEL:3

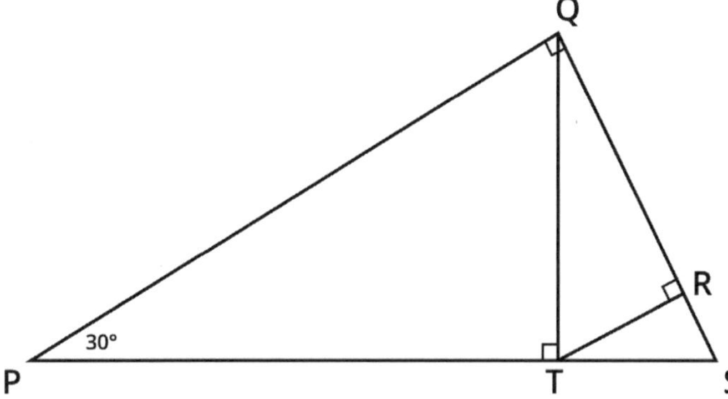

On the diagram above, if $PQ=1$, what is the length of RS?

Ⓐ $\frac{1}{12}$

Ⓑ $\frac{\sqrt{3}}{12}$

Ⓒ $\frac{1}{6}$

Ⓓ $\frac{\sqrt{2}}{12}$

163. LEVEL:3

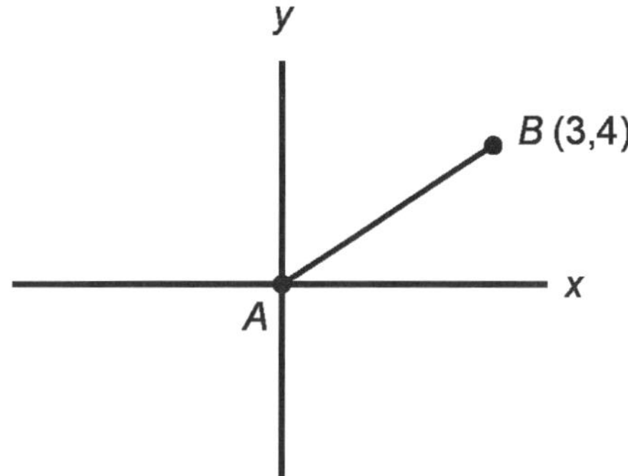

In the coordinate plane above, point C is not displayed. If the length of line segment BC is twice the length of segment AB, which of the following could not be the coordinates of point C?

Ⓐ $(-5, -2)$

Ⓑ $(9, 13)$

Ⓒ $(-3, -4)$

Ⓓ $(11, 10)$

164. LEVEL:3

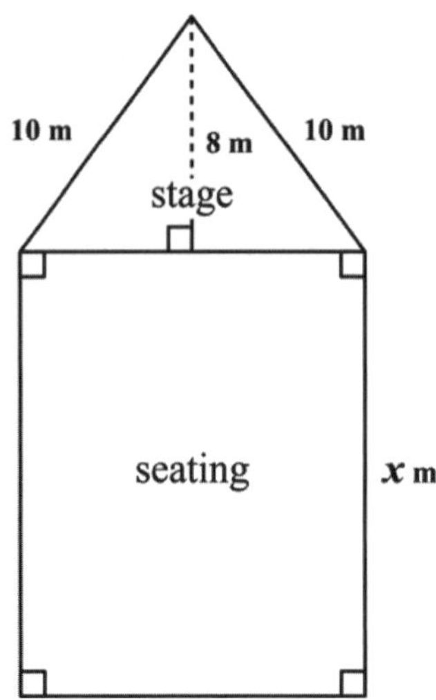

The figure above is the floor plan drawn by an architect for a small concert hall. The stage has depth 8 meters (m) and two walls each of length $10m$. If the seating portion of the hall has an area of 180 square meters, what is the value of x?

165. LEVEL:3

A plane flies approximately $30°$ north of east for 12 kilometers (km)and then continues for the next 31 km at approximately $60°$ north of east, as shown at left. What distance, d, in kilometers does the plane end up from its original location? (Round your answer to the nearest kilometer.)

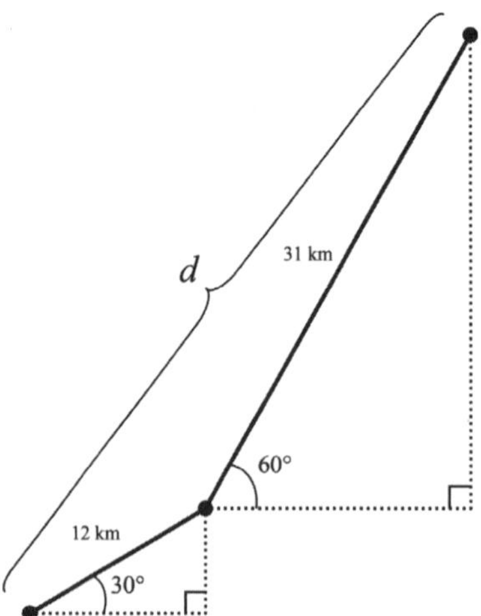

166. LEVEL:4

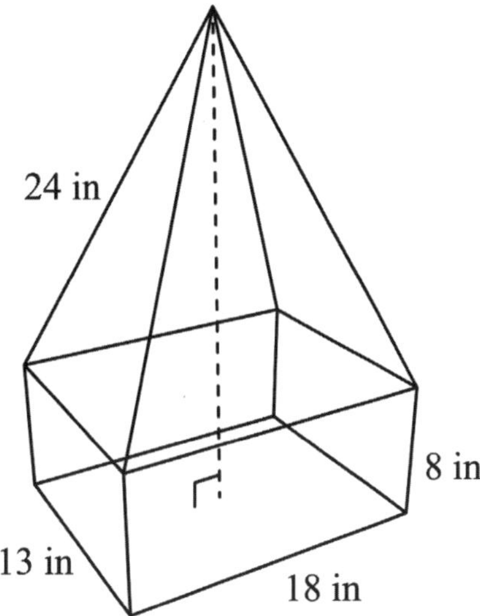

A hanging planter has the dimensions shown to the left in inches (in). Each support cable is of equal length. The pot for the planter is a right rectangular prism. About what is the total height (from where the cables meet to the base of the planter) of the hanging planter?

Ⓐ 21.28 in　Ⓑ 22.25 in　Ⓒ 29.28 in　Ⓓ 30.25 in

167. LEVEL:4

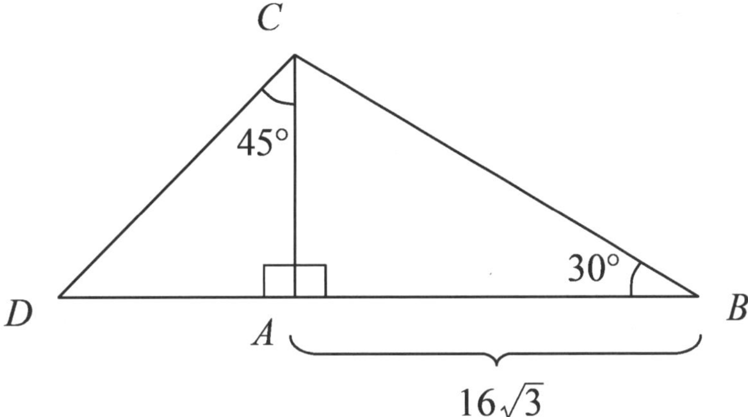

If $\overline{CD} = x\sqrt{2}$ in the figure shown above, What is the value of x ?

168.

LEVEL:4

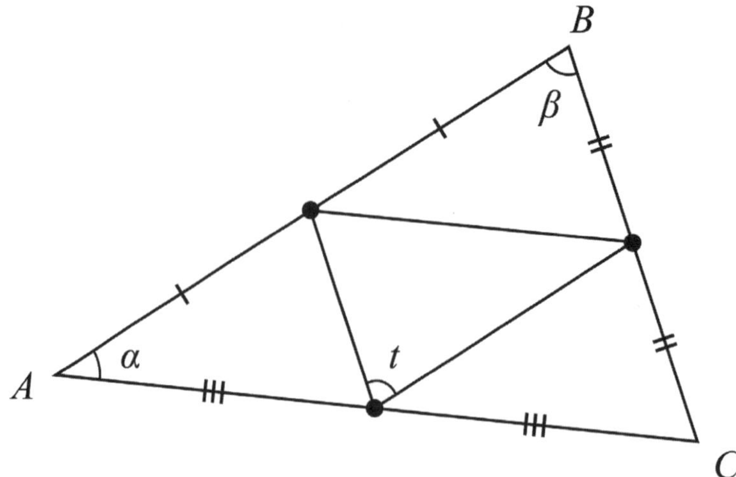

In triangle ABC shown at left, tick marks of equal number represent sides of equal length. Therefore, the point along line segment AB is the midpoint of AB. What is the value of β in terms of α and t?

Ⓐ a Ⓑ t Ⓒ $\frac{\pi}{2}-a$ Ⓓ $\frac{\pi}{2}-a$

169.

LEVEL:4

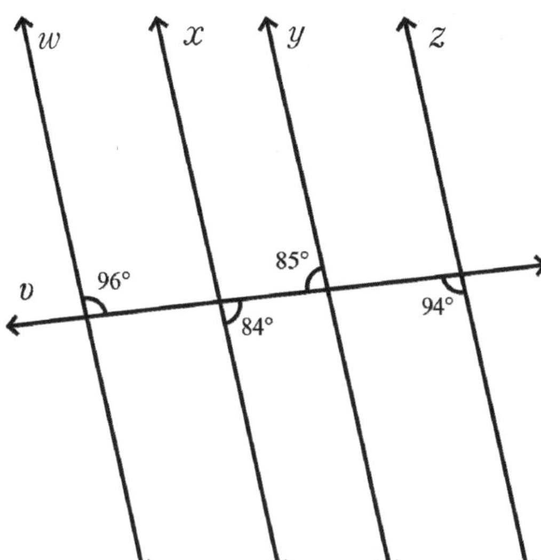

In the diagram at left, line v transverses lines w, x, y and z. Given the angle measurements shown, which pair of lines is parallel?

Ⓐ Lines w and x

Ⓑ Lines y and z

Ⓒ Lines w and y

Ⓓ Lines x and z

4.3 Right triangle trigonometry

1) Trig: Labeling Sides

In the right triangle below, the measure of one acute angle is labeled θ, and the sides of the triangle are labeled hypotenuse, opposite, and adjacent, according to their position relative to the angle of measure θ.

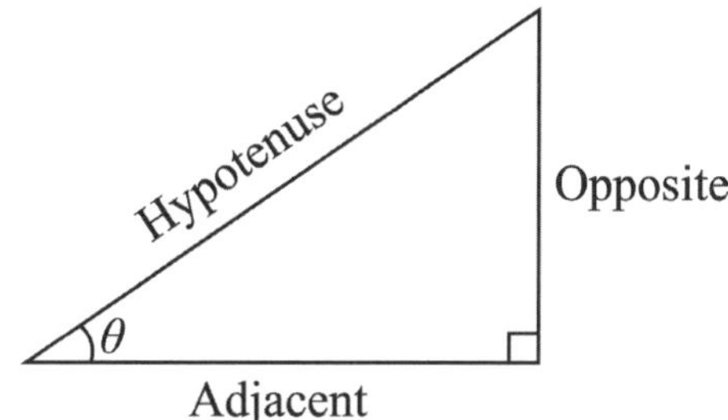

We label the three sides in this manner: The side opposite the right angle is called **the hypotenuse.** This side is labeled the same regardless of our choice of q. The labeling of the remaining two sides depend on our choice of theta; we therefore speak of these other two sides as being **adjacent to** the angle q or **opposite to** the angle q. The remaining side that touches the angle q is considered to be the side **adjacent to** q, and the remaining side that is far away from the angle q is considered to be **opposite to** the angle q, as shown in the picture.

2) Sine

The sine of an angle is the ratio of the side opposite the angle over the hypotenuse.

$$\sin\theta = \frac{opposite}{hypotenuse}$$

3) Cosine

The cosine of an angle is the ratio of the side adjacent the angle over the hypotenuse.

$$\cos\theta = \frac{adjacent}{hypotenuse}$$

4) Tangent

The sine of an angle is the ratio of the side opposite the angle over the side adjacent to the angle.

$$\tan\theta = \frac{opposite}{adjacent}$$

SOH → $\sin\theta = \frac{O}{H}$

CAH → $\cos\theta = \frac{A}{H}$

TOA → $\tan\theta = \frac{O}{A}$

170.

LEVEL:4

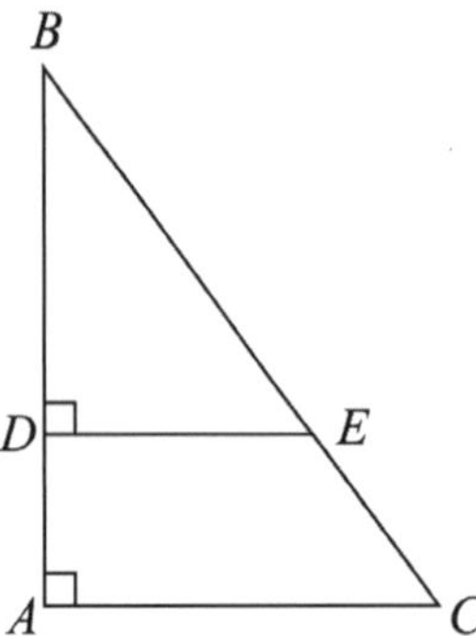

In the figure above, $\tan B = \frac{3}{4}$. If $BC = 15$ and $DA = 4$, what is the length of DE ?

171.

LEVEL:4

Suppose that the interior of a hinged car door forms a $38°$ angle with the body of the car when it is open. Given the measurements in the following figure, what is the width w, in inches, of the opening of the door? Round to the nearest inch.

Note: $\sin(19°) \approx 0.326$ and $\tan(19°) \approx 0.344$.

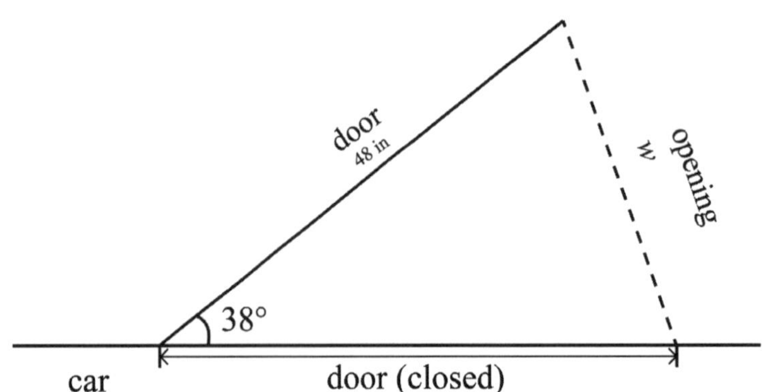

172. LEVEL:4

In the figure to the left, $\sin(x°)=0.9$. Which of the following is nearest to $\cos(90° - x°)$?

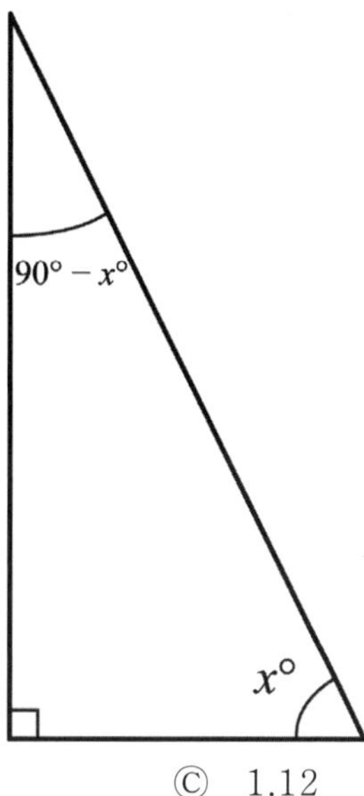

Ⓐ 0.45 Ⓑ 0.90 Ⓒ 1.12 Ⓓ 1.96

173. LEVEL:4

What is the value of x in the figure shown at left?

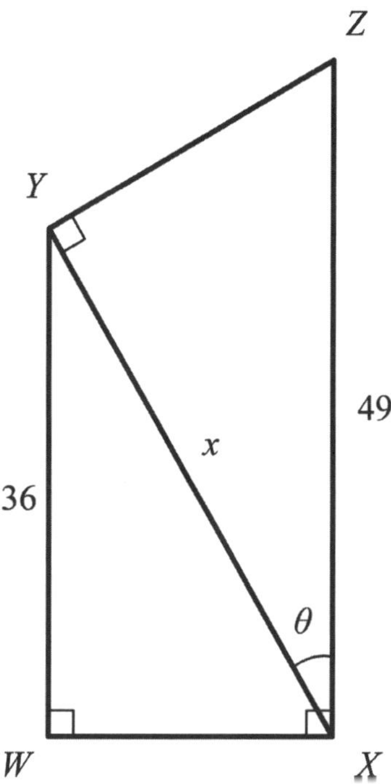

174. LEVEL:4

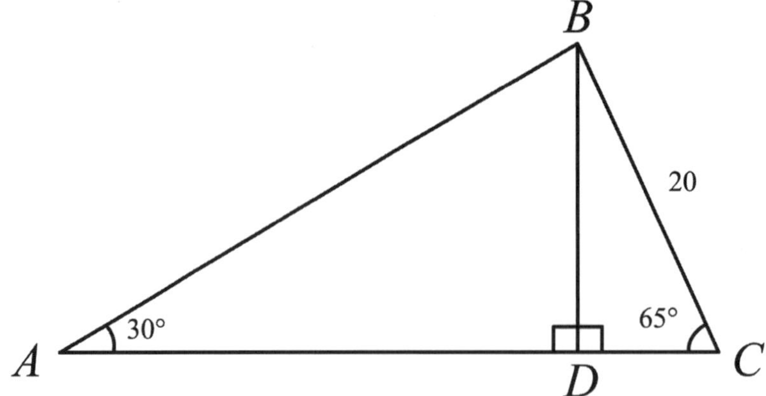

It is given that $\sqrt{3} \approx 1.732$. It is given that $\sin(65°) \approx 0.91$, $\cos(65°) \approx 0.42$, and $\tan(65°) \approx 2.14$. In the diagram the above, $\angle ADB$ and $\angle CDB$ are right angles. What is the length of AC to the nearest integer?

4.4 Angles, Arc lengths, Trigonometric functions

1. Angles Larger Than $90°$

Angles in a right triangle can never be larger than $90°$, since the sum of all three angles must equal $180°$. But on the Math IIC you may occasionally run into angles that are larger than $90°$. It is often more intuitive to think of these in terms of the coordinate plane rather than in terms of a triangle.

Below are pictured four angles in the coordinate plane. The first is the acute angle we've already covered in this chapter; the next three are all larger than $90°$.

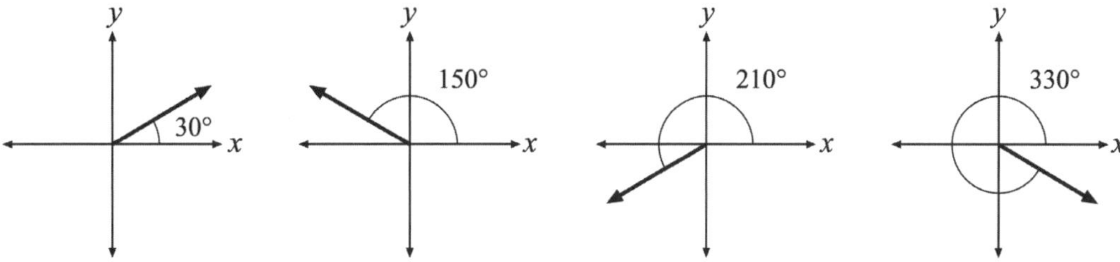

2. Radian measure

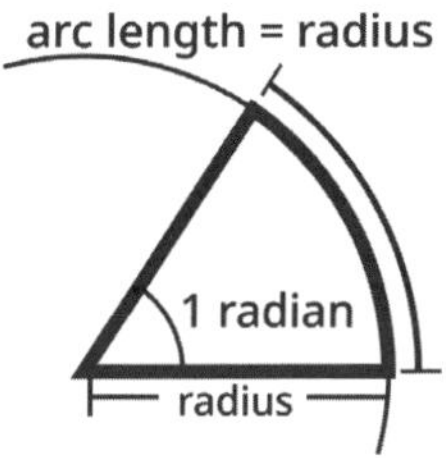

An angle of 1 radian subtends an arc equal in length to the radius of the circle.

One radian is the angle subtended at the center of a circle by an arc of circumference that is equal in length to the radius of the circle.

$$1\ radian = \frac{180°}{\pi},\ 1° = \frac{\pi}{180}\ radian$$

3. Arc Length and Sector Area

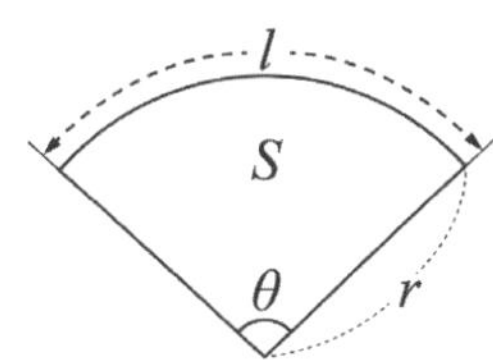

In diagram, θ is measured in **radians**

1) Arc Length : $l = r\theta$

2) Area of Sector : $S = \frac{1}{2}r^2\theta = \frac{1}{2}rl$

175. LEVEL:3

If $\theta = \frac{\pi}{2}$ radians, which of the following shows the measure of θ in degrees?

Ⓐ 45°

Ⓑ 90°

Ⓒ 135°

Ⓓ 180°

176. LEVEL:4

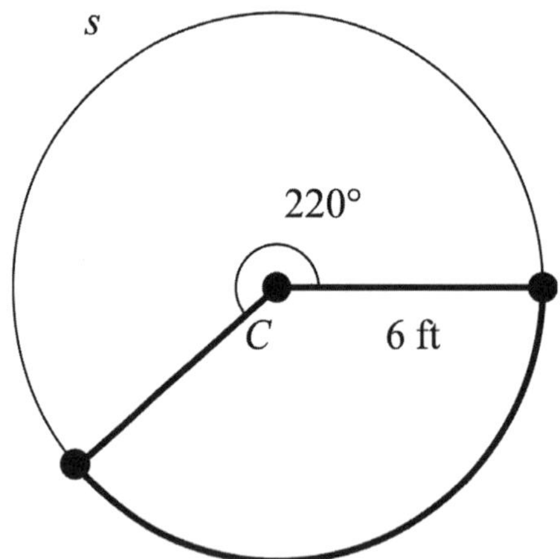

In the diagram at left, circle C has a radius of 6 feet (ft). Which of the following best approximates the measure of the arc length s?

Ⓐ $23ft$ Ⓑ $46ft$ Ⓒ $220ft$ Ⓓ $1320ft$

177. LEVEL:4

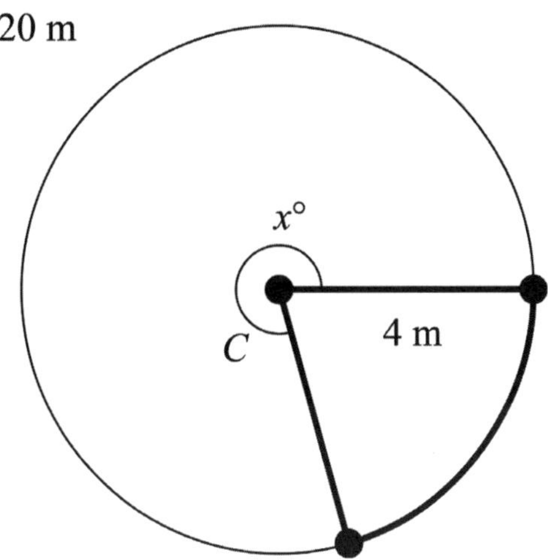

In a circle with center C and radius 4 meters (m), a central angle of $x°$ intercepts an arc of 20m as shown in the diagram at left. Rounded to the nearest degree, which of the following best approximates the value of x?

Ⓐ 5

Ⓑ 80

Ⓒ 286

Ⓓ 304

4.5 Circle theorems

1. What is a Circle?

A circle is the locus of all points equidistant from a central point.

2. Terms Related to Circles

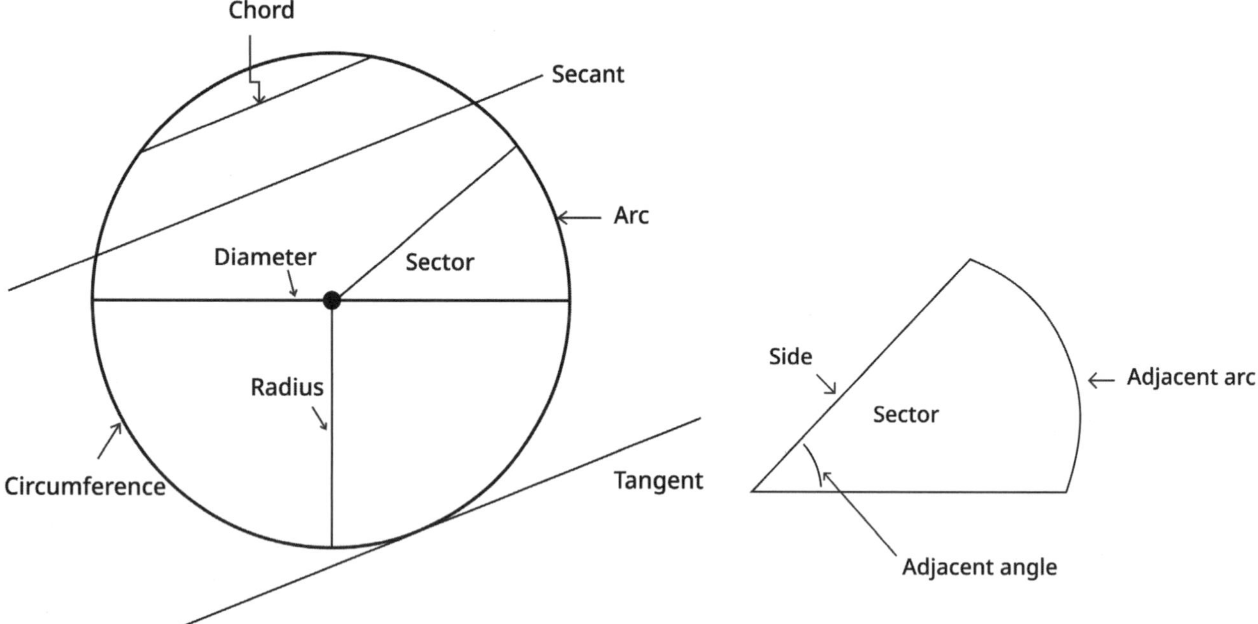

(1) Arc: a curved line that is part of the circumference of a circle

(2) Chord: a line segment within a circle that touches 2 points on the circle.

(3) Circumference: the distance around the circle.

(4) Diameter: the longest distance from one end of a circle to the other.

(5) Origin: the center of the circle

(6) Pi (π): A number, 3.141592⋯ equal to (the circumference) / (the diameter) of any circle.

(7) Radius: distance from center of circle to any point on it.

(8) Tangent of circle: a line perpendicular to the radius that touches ONLY one point on the circle.

(9) Sector: is like a slice of pie (a circle wedge).

3. Circle Formulas

If r is radius of a circle ann π = 3.141592⋯,

(1) Diameter = 2 x radius of circle = $2r$

(2) Circumference of Circle = π x diameter = 2 π x radius = $2\pi r$

(3) Area of Circle : $Area = \pi r^2$

(4) Length of a Circular Arc with central angle θ

- if the angle θ is in degrees, then length:

$$l = 2\pi r \times \frac{\theta}{360}$$

- if the angle θ is in radians, then length:

$$l = r\theta$$

(5) Area of Circle Sector with central angle θ

- if the angle θ is in degrees, then area:

$$A = \pi r^2 \times \frac{\theta}{360}$$

- if the angle θ is in radians, then area:

$$A = \frac{1}{2} r^2 \theta$$

4. Angle in a Circle

(1)

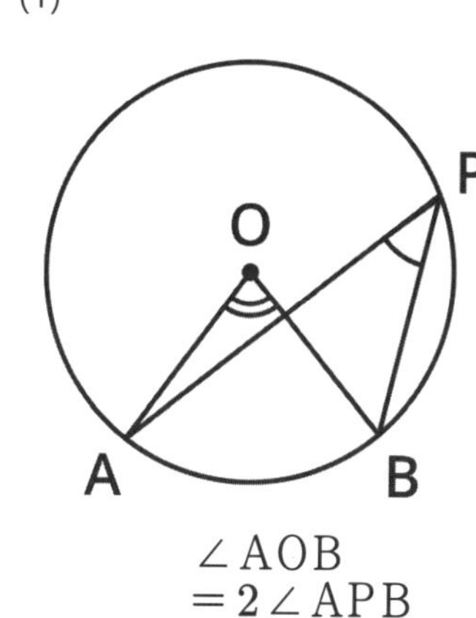

$\angle AOB = 2\angle APB$

(2)

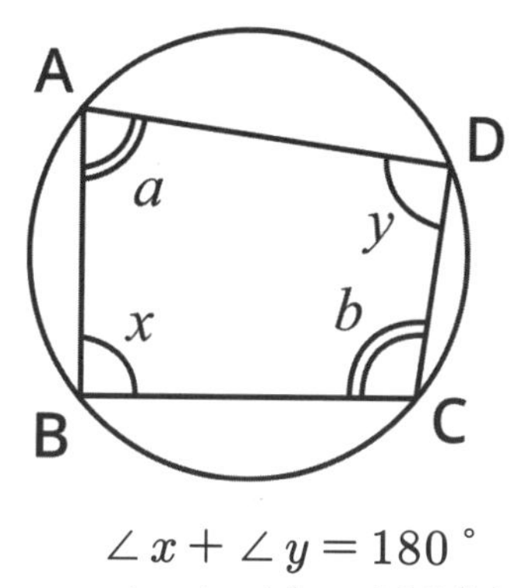

$\angle x + \angle y = 180°$
$\angle a + \angle b = 180°$

(3)

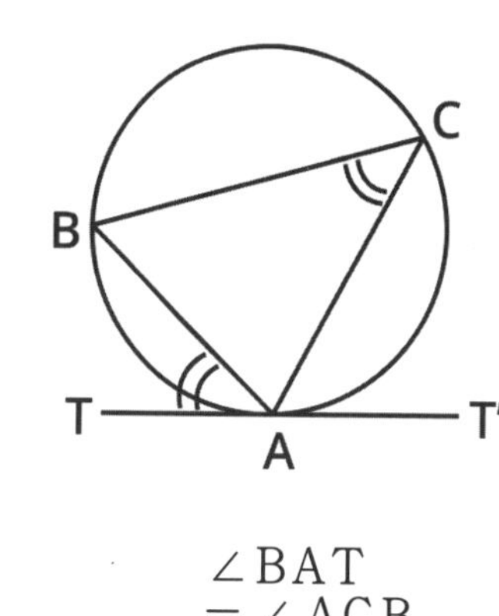

$\angle BAT = \angle ACB$

178.

LEVEL:4

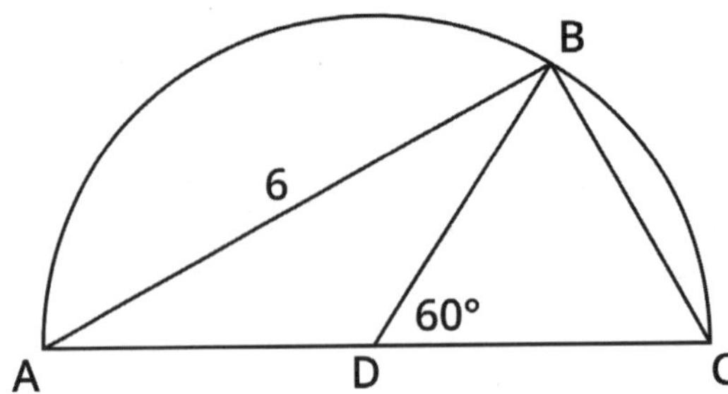

The triangle ABC is inscribed in a semicircle centerd at D. What is the area of triangle ABC?

Ⓐ $\dfrac{12}{\sqrt{3}}$

Ⓑ $6\sqrt{3}$

Ⓒ 12

Ⓓ $12\sqrt{3}$

179.

LEVEL:4

The circumference of Earth is estimated to be $40{,}030$ kilometers at the equator. Which of the following best approximates the diameter in miles, of Earth's equator? (1 kilometer ≈ 0.62137 miles)

Ⓐ $3{,}205$ miles

Ⓑ $5{,}541$ miles

Ⓒ $7{,}917$ miles

Ⓓ $13{,}004$ miles

180.

LEVEL:4

Maria gets new tires for her car. The radius of each of her old tires is 0.30 meter, and the radius of her old tires is 0.30 meter, and the radius of each of her new tires is 11% larger than the radius of one of her old tires. What is the circumference of each new tire, to the nearest tenth of a meter?

181.

LEVEL:4

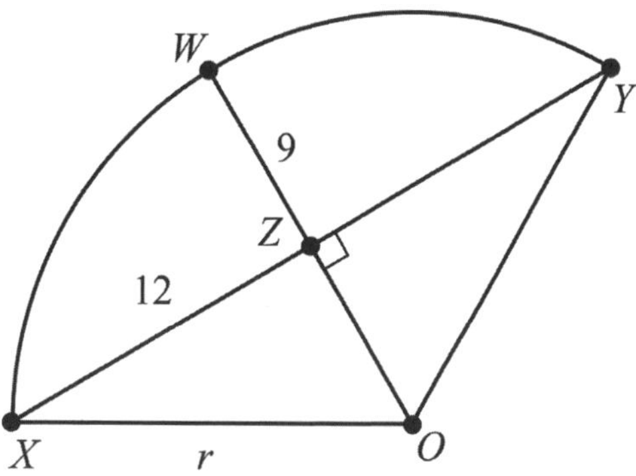

The sector of a circle shown above has center at O. Radius is perpendicular to chord and intersects at point Z. The length of is 9 and the length of is 12. To the nearest tenth, what is the radius, r, of the circle?

182. LEVEL:4

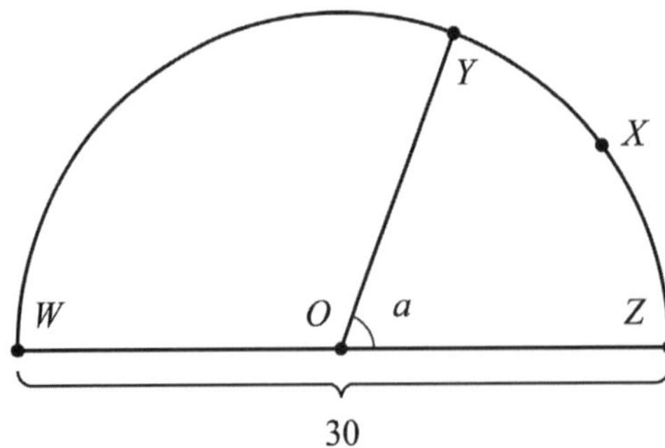

The semicircle shown above has its center at point O. The diameter of the circle is 30, and the arc YXZ has length 18. To the nearest hundredth of a radian, what is the measure, a, of angle YOZ ?

183. LEVEL:4

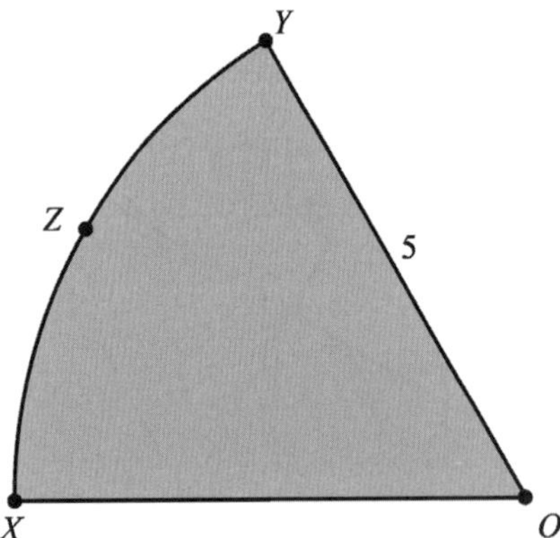

The sector of a circle shown above has center at O. The radius of the circle has length 5. The arc XZY has length 7. To the nearest tenth, what is the shaded area of the sector?

184.

LEVEL:4

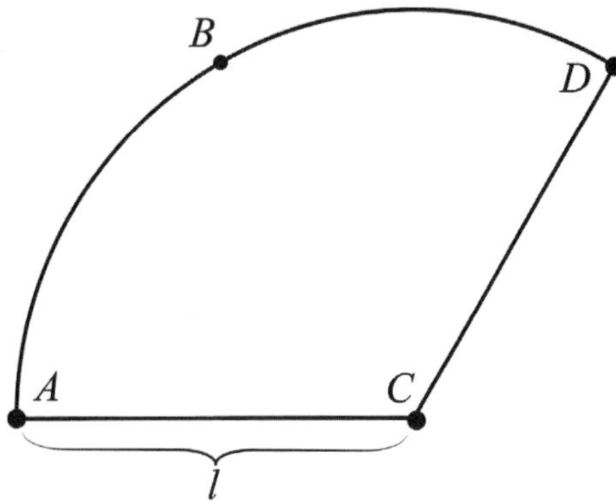

The sector of a circle shown at left has center C. The total area of the sector is 500. In addition, arc ABD has length 40. What is the length, l, of line segment?

Ⓐ 12.5

Ⓑ 16

Ⓒ 20

Ⓓ 25

185.

LEVEL:5

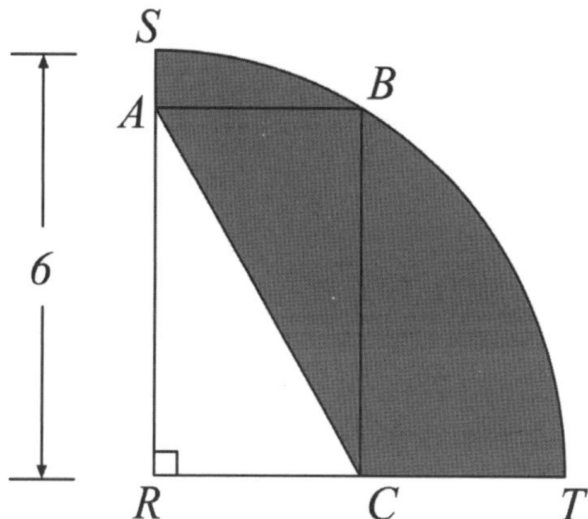

In the figure, arc SBT is one quarter of a circle with center R and radius 6. If the length plus the width of rectangle $ABCR$ is 8. then what is the perimeter of the shaded region?. Find the closest value.

Ⓐ $8+3\pi$

Ⓑ $10+3\pi$

Ⓒ $12+3\pi$

Ⓓ $14+3\pi$

4.6 Circle equation

1. standard form:

$$(x-h)^2+(y-k)^2=r^2$$

where (h,k) is the center of the circle and r is the radius. When the circle is centered at the origin, so that $h=k=0$, then the equation simplifies to:

$$x^2+y^2=r^2$$

2. General Form

The general form of the equation of a Equation of Circle is:

$$x^2+y^2+Ax+By+C=0$$

1) center of the circle : $\left(-\frac{A}{2}, -\frac{B}{2}\right)$

2) Length of radius : $\frac{\sqrt{A^2+B^2-4C}}{2}$

186. LEVEL:3

A circle in the xy-plane has equation $(x+3)^2+(y-1)^2=25$. which of the following points does lie in the interior of the circle?

Ⓐ $(-7, 3)$

Ⓑ $(3, -7)$

Ⓒ $(7, -3)$

Ⓓ $(-7, -3)$

187. LEVEL:4

A circle in the xy-plane has the equation:

$$2x^2+2y^2-8x-6y-16=0$$

What is the diameter of the circle? Round to the nearst whole number.

188. LEVEL:4

A circle in the $x^2+y^2-6x-10y=2$. What is the value of $a+b$ if the equation has the center point (a,b) of the circle?

189. LEVEL:4

A circle in the xy−plane has a center at $(\frac{5}{8}, -\frac{5}{6})$ and a diameter of $\frac{7}{10}$, Which of the following is an equation of the circle?

Ⓐ $\left(x+\frac{5}{8}\right)^2+\left(y-\frac{6}{5}\right)^2=\frac{49}{100}$

Ⓑ $\left(x+\frac{5}{8}\right)^2+\left(y-\frac{6}{5}\right)^2=\frac{49}{400}$

Ⓒ $\left(x-\frac{5}{8}\right)^2+\left(y+\frac{6}{5}\right)^2=\frac{49}{100}$

Ⓓ $\left(x-\frac{5}{8}\right)^2+\left(y+\frac{6}{5}\right)=\frac{49}{400}$

190. LEVEL:4

The diameter of a circle graphed in the xy-plane has endpoints at $(-23, 15)$ and $(1, -55)$. Which of the following is an equation of the circle?

Ⓐ $(x+23)^2+(y-15)^2=1369$

Ⓑ $(x+23)^2+(y-15)^2=5476$

Ⓒ $(x+11)^2+(y+20)^2=5476$

Ⓓ $(x+11)^2+(y+20)^2=1369$

191. LEVEL:4

A circle in the xy-plane has the equation $x^2+y^2-10x+34y-527=0$. If the y-coordinate of a point on the circle is -38, what is a possible x-coordinate?

4.7 Trigonometric Definitions on xy Plane and Basic Properties

1. Trigonometric Definition on the Circle with center $(0, 0)$

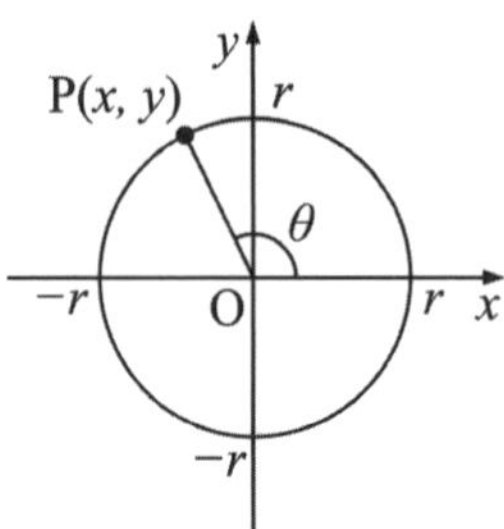

1) $\sin\theta = \frac{y}{r}$, $\cos\theta = \frac{x}{r}$, $\tan\theta = \frac{y}{x}$,

2) $\text{cosec}\theta = \frac{r}{y}$, $\sec\theta = \frac{r}{x}$, $\cot\theta = \frac{x}{y}$,

2. Trigonometric $(+/-)$ Sign on Each Quadrants

The four quadrants of the coordinate plane become very important when dealing with angles that are larger than $90°$.

1) On the 1st quadrant : Every value is $+$.
2) On the 2nd quadrant : Only $\sin$ value is $+$.
3) On the 3rd quadrant : Only $\tan$ value is $+$.
4) On the 4nd quadrant : Only $\cos$value is $+$.

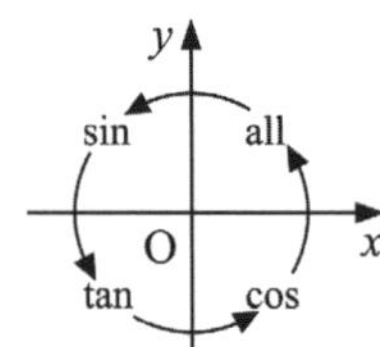

All **S**illy **T**urtle **C**rawl.

All **S**tudent can **T**ake **C**alculus.

3. Basic Rules(Properties) for Trigonometry

1) $\text{cosec}\theta = \frac{1}{\sin\theta}$, $\sec\theta = \frac{1}{\cos\theta}$, $\cot\theta = \frac{1}{\tan\theta}$

2) Tangent ratio : $\tan\theta = \frac{\sin\theta}{\cos\theta}$, $\cot\theta = \frac{\cos\theta}{\sin\theta}$

3) $\sin^2\theta + \cos^2\theta = 1$, $1 + \tan^2\theta = \sec^2\theta$

$1 + \cot^2\theta = \text{cosec}^2\theta$

192. LEVEL:3

Which of the following is the value of $\sin\left(\frac{\pi}{2}\right)$?

Ⓐ -1

Ⓑ 0

Ⓒ $\frac{\sqrt{2}}{2}$

Ⓓ 1

193. LEVEL:3

In xy-plane, one angle measures x°, where $\sin x^\circ = \frac{4}{5}$. What is the $\cos(90^\circ - x^\circ)$?

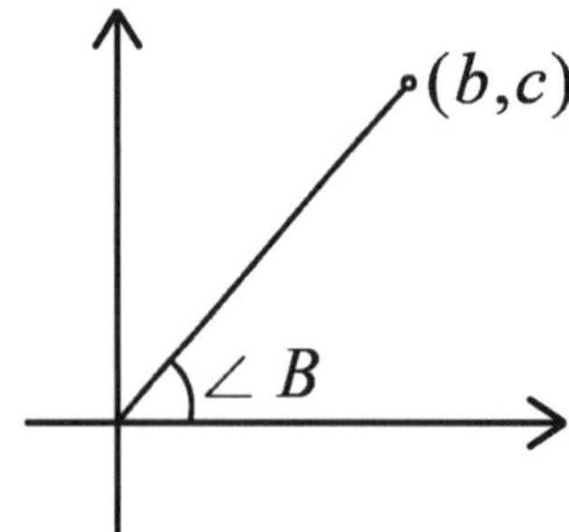

Given a point (b, c) on xy plane above, which of the following is equal to $\frac{c}{b}$?

Ⓐ $\tan B$

Ⓑ $\frac{1}{\tan B}$

Ⓒ $\cos B$

Ⓓ $\frac{1}{\cos B}$

4.8 Complex numbers

1. Definition of Unit Imaginary Number "i'

The imaginary is defined to be

1) $i = \sqrt{-1}$
2) $i^2 = (\sqrt{-1})^2 = -1$
3) Pattern of powers :

$i^1 = i$
$i^2 = -1$
$i^3 = -i$
$i^4 = 1$
$i^5 = i^1 = i$
$i^6 = i^2 = -1$
$i^7 = i^3 = -i$
$i^8 = i^4 = 1$

$\Rightarrow$

$i^{4n} = 1$
$i^{4n+1} = i$
$i^{4n+2} = -1$
$i^{4n+3} = -i$

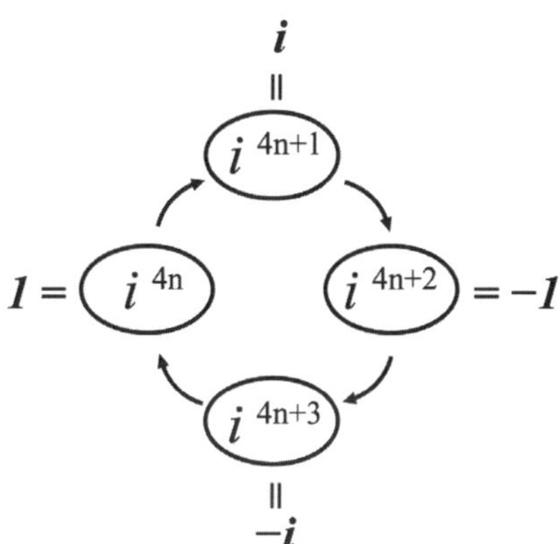

2. General form of Complex Number

if a and b are real numbers $z = a + bi$

1) If $a \neq 0$, $b = 0$ then z is real number.
2) If $a = 0$, $b \neq 0$ then z is purely imaginary number.
3) If $a \neq 0$, $b \neq 0$ then z is non−purely imaginary number.
4) Conjugates : $\bar{z} = a + bi$

3. Operations on Complex Numbers

If a, b, c and d are real numbers

1) $(a+bi)+(c+di) = (a+c)+(b+d)i$
2) $(a-bi)+(c-di) = (a-c)+(b-d)i$
3) $(a+bi)(c+di) = (ac-bd)+(ad+bc)i$
4) $\dfrac{a+bi}{c+di} = \dfrac{ac+bd}{c^2+d^2} + \dfrac{bc-ad}{c^2+d^2}$ $(c+di \neq 0)$

4. Modulus of Complex Numbers

1) The modulus of the complex number $z = a + bi$ is the real number $|z| = \sqrt{a^2 + b^2}$
2) The modulus of the complex number $z = a + bi$ is the length of the corresponding vector $\begin{pmatrix} a \\ b \end{pmatrix}$

195. LEVEL:3

$$i^3+i^2$$

Which of the following is equivalent to the complex number shown above?

Note: $i=\sqrt{-1}$

Ⓐ -1

Ⓑ -2

Ⓒ $-1+i$

Ⓓ $-1-i$

196. LEVEL:3

Simplify : $\sqrt{-3}+\sqrt{-9}+\sqrt{-16}$

Ⓐ $i\sqrt{3}-7i$

Ⓑ $i\sqrt{3}-i$

Ⓒ $i\sqrt{3}+7i$

Ⓓ $i\sqrt{3}+i$

197.

LEVEL:4

$$m^2+6m+10=0$$

Which of the following are solutions to the equation above?

I. $-3+i$

II. $-3-i$

III. $3+i$

Note: $i=\sqrt{-1}$

Ⓐ I only

Ⓑ I and II only

Ⓒ I and III only

Ⓓ I, II, and III

198.

LEVEL:4

$$704i^{1776}$$

Which of the following is equivalent to the complex number shown above?

Note: $i=\sqrt{-1}$

Ⓐ 704

Ⓑ −704

Ⓒ 704i

Ⓓ −704i

199. LEVEL:4

$$\frac{5+7i}{6-3i}$$

Which of the following is equivalent to the complex number shown above?

Note: $i=\sqrt{-1}$

Ⓐ $\dfrac{9+57i}{45}$

Ⓑ $\dfrac{9+57i}{3}$

Ⓒ $\dfrac{51+57i}{45}$

Ⓓ $\dfrac{51+57i}{3}$

200 LEVEL:4

$$\frac{1}{2+5i}-\frac{4+3i}{3-i}$$

Which of the following is equivalent to the complex number shown above?

Note: $i=\sqrt{-1}$

Ⓐ $\dfrac{10+26i}{(2+5i)(3-i)}$

Ⓑ $\dfrac{10-26i}{(2+5i)(3-i)}$

Ⓒ $\dfrac{10+27i}{(2+5i)(3-i)}$

Ⓓ $\dfrac{10-27i}{(2+5i)(3-i)}$

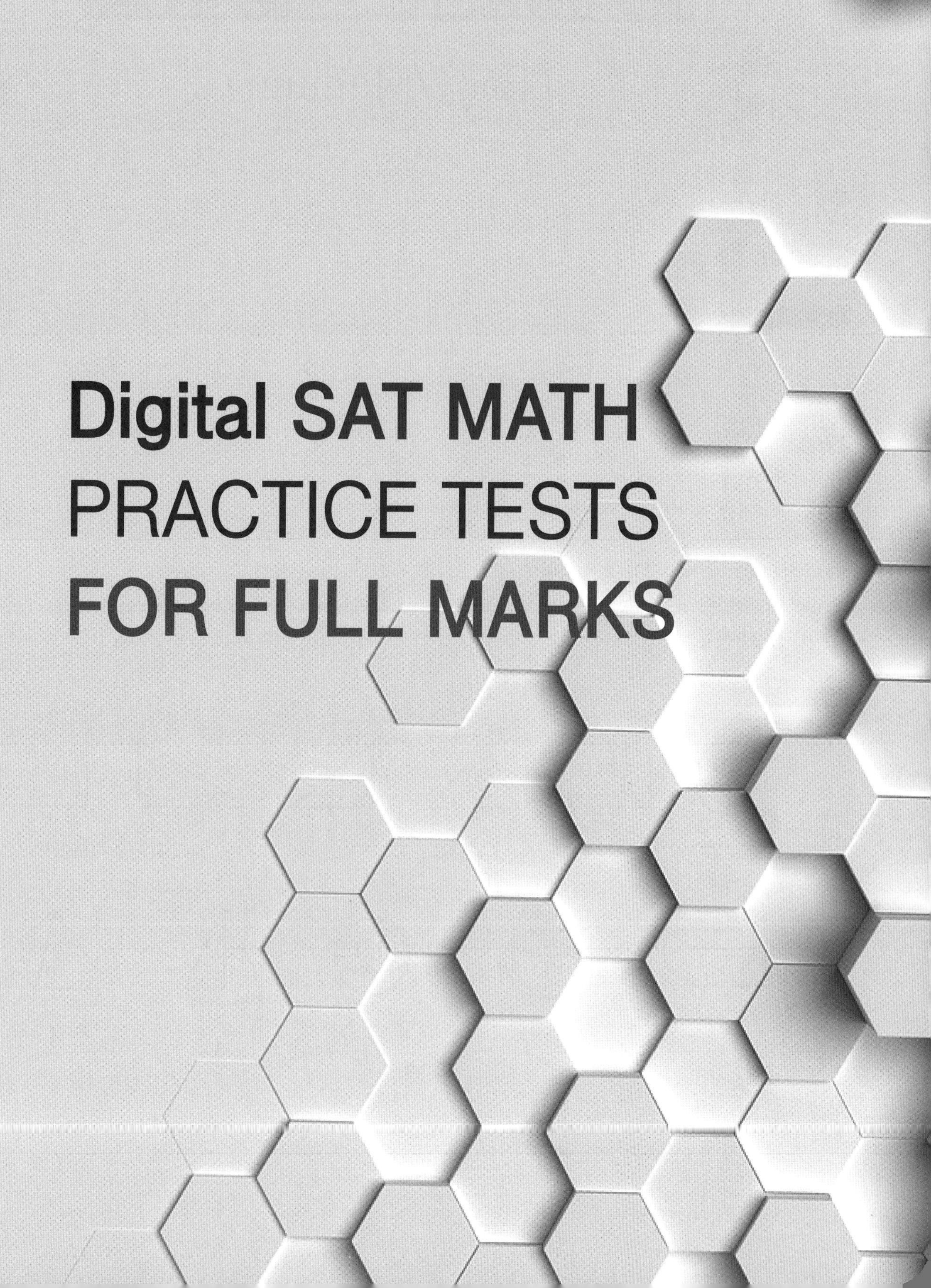

Digital SAT MATH PRACTICE TESTS FOR FULL MARKS

Final Practice Test 1

(Easy/Medium)

Math

22 Questions

DIRECTIONS

The questions in this section address a number of important math skills.
Use of a calculator is permitted for all questions.

NOTES

Unless otherwise indicated:

- All variables and expressions represent real numbers.
- Figures provided are drawn to scale.
- All figures lie in a plane.
- The domain of a given function fis the set of all real numbers x for which f(x) is a real number.

REFERENCE

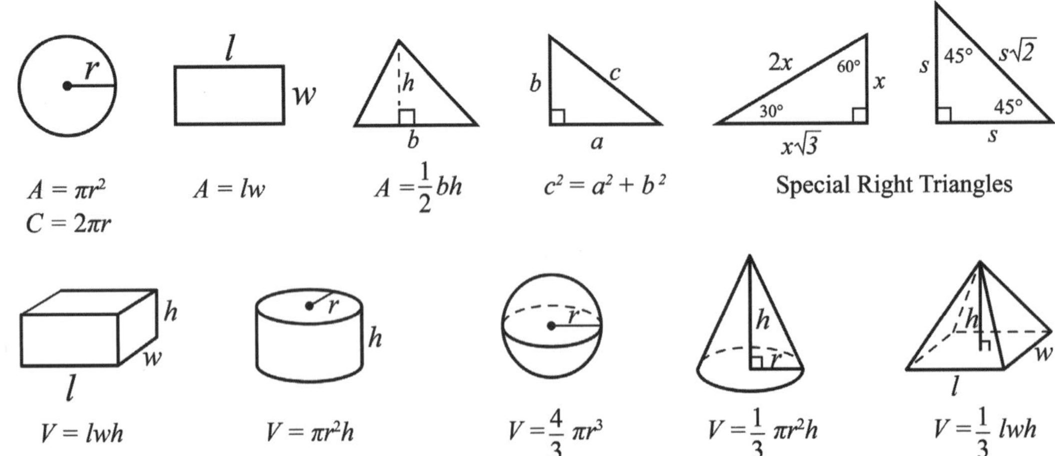

The number of degrees of arc in a circle is 360.
The number of radians of arc in a circle is 2π.
The sum of the measures in degrees of the angles of a triangle is 180.

For multiple-choice questions, solve each problem, choose the correct answer from the choices provided, and then circle your answer in this book.
Circle only one answer for each question. If you change your mind, completely erase the circle. You will not get credit for questions with more than one answer circled, or for questions with no answers circled.

For student-produced response questions, solve each problem and write your answer next to or under the question in the test book as described below.

- Once you've written your answer, circle it clearly. You will not receive credit for anything written outside the circle, or for any questions with more than one circled answer.

- If you find more than one correct answer, write and circle only one answer.

- Your answer can be up to 5 characters for a positive answer and up to 6 characters (including the negative sign) for a negative answer, but no more.

- If your answer is a fraction that is too long (over 5 characters for positive, 6 characters for negative), write the decimal equivalent.

- If your answer is a decimal that is too long (over 5 characters for positive, 6 characters for negative), truncate it or round at the fourth digit.

- If your answer is a mixed number (such as $3\frac{1}{2}$), write it as an improper fraction ($7/2$) or its decimal equivalent (3.5).

- Don't include symbols such as a percent sign, comma, or dollar sign in your circled answer.

1.

$$4x+1=-ax-4$$

In the equation shown above, a is a constant. Which of the following values of a results in an equation with exactly one solution?

Ⓐ 4

Ⓑ -4

Ⓒ 4 or -4

Ⓓ Neither value

2.

Two lines graphed in the xy-plane have the equations $2x+5y=20$ and $y=kx-3$, where k is a constant. For what value of k will the two lines be perpendicular?

3.

Three kids own a total of 96 comic books. If one of the kids owns 16 of the comic books, what is the average (arithmetic mean) number of comic books owned by the other two kids?

Ⓐ 40

Ⓑ 42

Ⓒ 44

Ⓓ 46

4.

A bird traveled 72 miles in 6 hours flying at constant speed. At this rate, how many miles did the bird travel in 5 hours?

Ⓐ 12

Ⓑ 30

Ⓒ 60

Ⓓ 66

5.

Franklin bought several kites, each costing 16 dollars. Richard purchased several different kites, each costing 20 dollars. If the ratio of the number of kites Franklin purchased to the number of kites Richard purchased was 3 to 2, what was the average cost of each kite they purchased?

Ⓐ \$16.80

Ⓑ \$17.20

Ⓒ \$17.60

Ⓓ \$18.00

6.

A specialized machine can place a line of cones along the highway at a rate of 30 cones per minute. The cones are spaced an average of 15 meters apart. Which of the following equations could be used to describe the total distance in meters, d , lined by the cones as a function of t, the time in minutes?

Ⓐ $d=2t$

Ⓑ $d=15t+30$

Ⓒ $d=30t+15$

Ⓓ $d=450t$

7.

$$16x^2 - 8x - 3 = 0$$

Let $x = q$ and $x = r$ be solutions to the equation shown above, with $q > r$. What is the value of $q - r$?

8.

$$\sqrt{0.05} \times \sqrt{15}$$

Which of the following values is equal to the value above?

Ⓐ $\dfrac{\sqrt{3}}{2}$

Ⓑ $\dfrac{3}{4}$

Ⓒ $\sqrt[4]{0.75}$

Ⓓ 0.375

9.

Circle A is inside Circle B, and the two circles share the same center O. If the circumference of B is four times the circumference of A, and the radius of Circle A is three, what is the difference between Circle B's diameter and Circle A's diameter?

Ⓐ 6

Ⓑ 9

Ⓒ 12

Ⓓ 18

10.

Under which conditions is the expression $\frac{ab}{a-b}$ always less than zero?

Ⓐ $a < b < 0$

Ⓑ $0 < b < a$

Ⓒ $a < 0 < b$

Ⓓ $b < a < 0$

11.

A sphere with a diameter of 5 cm is being machined out of a cube that is just large enough to contain the inscribed sphere. If the sphere weighs 52 grams, which of the following best approximates the weight of material removed from the cube during the machining of the sphere?

Note: The volume of a sphere $= \frac{4}{3}\pi r^3$

Ⓐ $75g$

Ⓑ $60g$

Ⓒ $54g$

Ⓓ $47g$

12.

Line F can be described by the function $f(x)=5x$. Line G is parallel to Line F such that the shortest distance between Line G and Line F is c, and the y-intercept of Line G is negative. Which of the following is a possible equation for line G?

Ⓐ $g(x)=x-5$

Ⓑ $g(x)+5\sqrt{2}=5x$

Ⓒ $g(x)=x-5\sqrt{2}$

Ⓓ $g(x)-5=5x$

13.

What is one possible value of $x+2$ for the equation below?

$$\frac{1}{3x+6}=\frac{x+2}{48}$$

Ⓐ -6

Ⓑ -2

Ⓒ 2

Ⓓ 4

14.

Parallelogram $QRST$ has an area of 120 and its longest side (QT) is 24. The angle opposite the vertical is 30°, and the vertical is from R to point U, which lies along QT. What is the length of the hypotenuse of Triangle QRU, rounded to the nearest whole number?

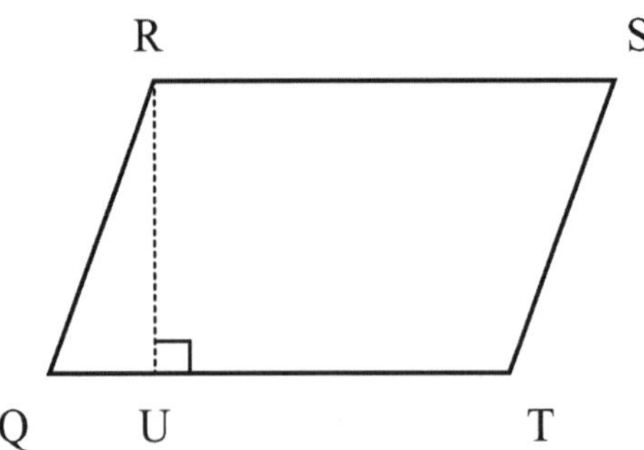

Ⓐ 6

Ⓑ 8

Ⓒ 10

Ⓓ 13

15.

A movie theater charges $11.50 for an Adult ticket and $9.75 for a Child ticket. The theater offers a 20% discount on all tickets are purchased together. Brenda spent $95.60 total buying tickets for her group. Seven (7) of the tickets she purchased were Adult tickets. How many Child tickets did she purchase?

Ⓐ 1

Ⓑ 3

Ⓒ 4

Ⓓ 5

16.

The first term of a sequence is the number n, and each term thereafter is 5 greater than the term before. Which of the following is the average (arithmetic mean) of the first nine terms of this sequence?

Ⓐ $n+20$

Ⓑ $n+180$

Ⓒ $2n$

Ⓓ $2n+40$

17.

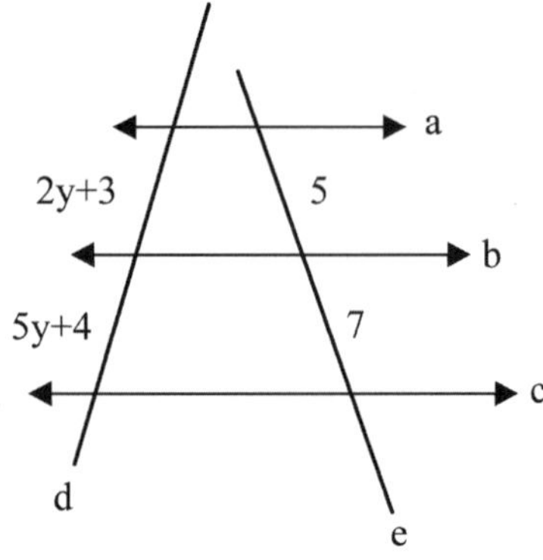

What is the value of y, if lines a, b and c are parallel?

Ⓐ $\frac{41}{12}$

Ⓑ $\frac{41}{11}$

Ⓒ $\frac{31}{9}$

Ⓓ $\frac{31}{7}$

18.

The circumference of a right cylinder is half its height. The radius of the cylinder is x. What is the volume of the cylinder in terms of x?

Ⓐ $2\pi x^3$

Ⓑ $3\pi x^3$

Ⓒ $3\pi^2 x^3$

Ⓓ $4\pi^2 x^3$

19.

What is the sum of the areas of the three rectangles that are drawn below the graph of the line $y=2^x$.

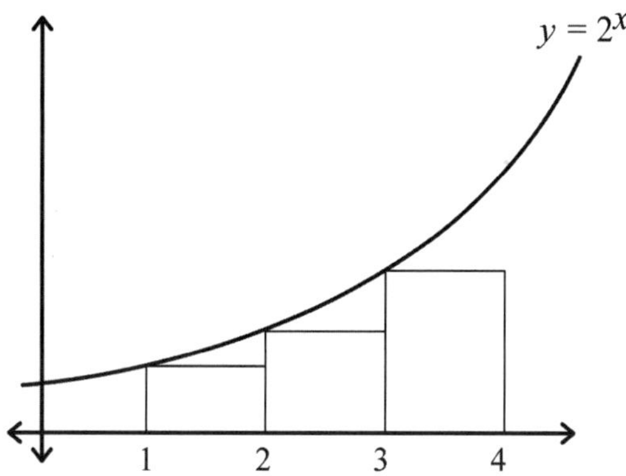

Ⓐ 6
Ⓑ 12
Ⓒ 14
Ⓓ 16

20.

Line F is represented by the equation $y=x+1$. Line G is represented by the equation $y=px+q$. Line F and Line G always intersects where $x=1$. What equation properly expresses the relationship between p and q ?

Ⓐ $p=-q+2$
Ⓑ $p=q-2$
Ⓒ $p=2q+1$
Ⓓ $p=2q-1$

21.

If AB is parallel to DC, and AD is parallel to BC, then what is the value of $b-a$?

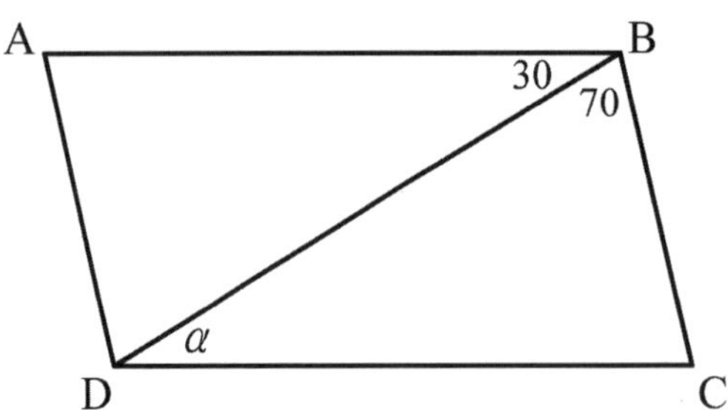

Ⓐ 30°
Ⓑ 50°
Ⓒ 60°
Ⓓ 70°

22.

For triangle ABC shown below, what is its area rounded to the nearest tenth?

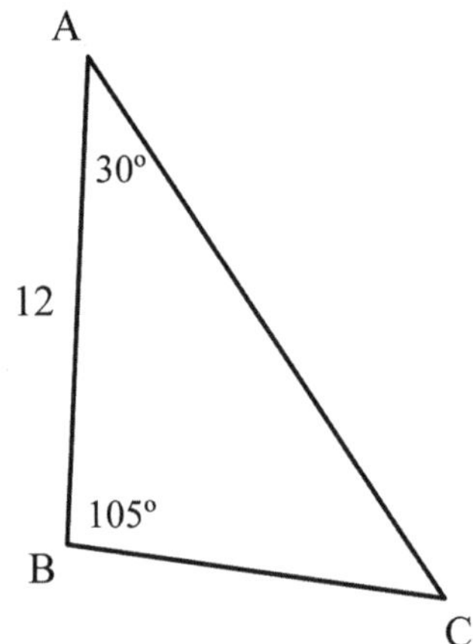

Ⓐ 49.2
Ⓑ 36.0
Ⓒ 41.6
Ⓓ 52.0

Final Practice Test 1

(Advanced/Difficult)

Math

22 Questions

DIRECTIONS

The questions in this section address a number of important math skills.
Use of a calculator is permitted for all questions.

NOTES

Unless otherwise indicated:

- All variables and expressions represent real numbers.
- Figures provided are drawn to scale.
- All figures lie in a plane.
- The domain of a given function fis the set of all real numbers x for which f(x) is a real number.

REFERENCE

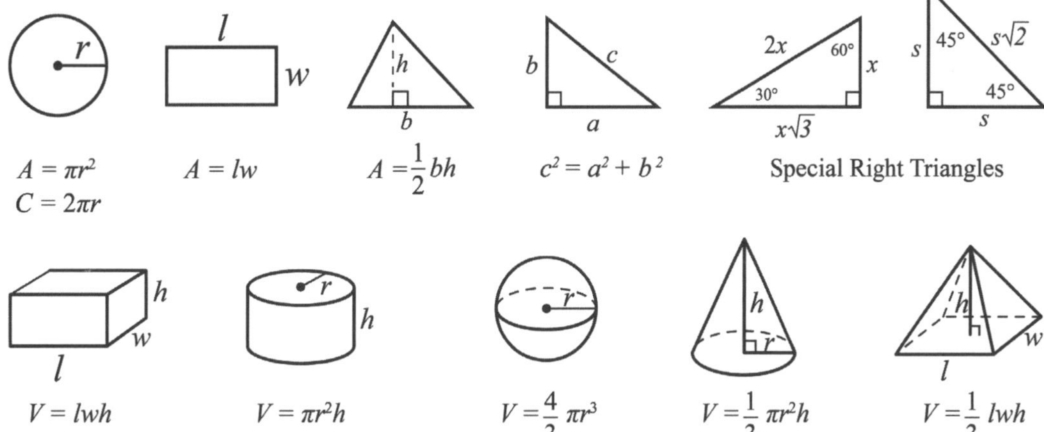

The number of degrees of arc in a circle is 360.

The number of radians of arc in a circle is 2π.

The sum of the measures in degrees of the angles of a triangle is 180.

For multiple-choice questions, solve each problem, choose the correct answer from the choices provided, and then circle your answer in this book.
Circle only one answer for each question. If you change your mind, completely erase the circle. You will not get credit for questions with more than one answer circled, or for questions with no answers circled.

For student-produced response questions, solve each problem and write your answer next to or under the question in the test book as described below.

- Once you've written your answer, circle it clearly. You will not receive credit for anything written outside the circle, or for any questions with more than one circled answer.

- If you find more than one correct answer, write and circle only one answer.

- Your answer can be up to 5 characters for a positive answer and up to 6 characters (including the negative sign) for a negative answer, but no more.

- If your answer is a fraction that is too long (over 5 characters for positive, 6 characters for negative), write the decimal equivalent.

- If your answer is a decimal that is too long (over 5 characters for positive, 6 characters for negative), truncate it or round at the fourth digit.

- If your answer is a mixed number (such as $3\frac{1}{2}$), write it as an improper fraction ($7/2$) or its decimal equivalent (3.5).

- Don't include symbols such as a percent sign, comma, or dollar sign in your circled answer.

1.

A passenger ship left Southampton, England for the Moroccan coast. The ship travelled the first 230 miles at an average speed of 20 knots, then increased its speed for the next 345 miles to 30 knots. It travelled the remaining 598 miles at an average speed of 40 knots. What was the ship's approximate average speed in miles per hour? (1 knot = 1.15 miles per hour)

Ⓐ 36

Ⓑ 38

Ⓒ 40

Ⓓ 42

2.

The initial number of elements in Set A is x, where $x > 0$. If the number of elements in Set A doubles every hour, which of the following represents the increase in the number of elements in Set A after exactly one day?

Ⓐ $23x^{23}$

Ⓑ $2^{23}x$

Ⓒ $23x^{24}$

Ⓓ $2^{24}x - x$

3.

$$5^{\frac{1}{3}} - 5^{\frac{4}{3}}$$

Which of the following expressions is equivalent to the expression above?

Ⓐ $-4 \times 5^{\frac{1}{3}}$

Ⓑ $-\sqrt[3]{620}$

Ⓒ $5^{\frac{1}{4}}$

Ⓓ $\frac{1}{5}$

4.

The midpoints of the sides of a square are connected to form a new inscribed square. How many times greater than the area of the inscribed square is the area of the original square?

Ⓐ $\frac{1}{2}$

Ⓑ 2

Ⓒ 8

Ⓓ $8\sqrt{2}$

5.

On a number line with points LMNOP, the ratio of LM to MN is 1, and the ratio of NO to OP is 3/4. If the length of LP is 28 and the length of MO is 13, what is the ratio of LO to MP?

Ⓐ $\frac{7}{18}$

Ⓑ $\frac{13}{28}$

Ⓒ $\frac{15}{28}$

Ⓓ $\frac{20}{21}$

6.

New York City Workforce

	Employed	Unemployed	Total
Men	22,000		
Women			21,500
Toral	40,000		45,500

The table above, which describes the New York City workforce, is only partially filled. Based on this information, what proportion of the total New York City workforce are unemployed women?

Ⓐ $\frac{43}{91}$

Ⓑ $\frac{7}{43}$

Ⓒ $\frac{11}{91}$

Ⓓ $\frac{1}{13}$

7.

Triangle ABC is the isosceles triangle shown in the illustration below. $\angle A = 70°$ and $\angle D = 45°$. Points A, B and D are collinear. Find x, the measure of $\angle BCD$.

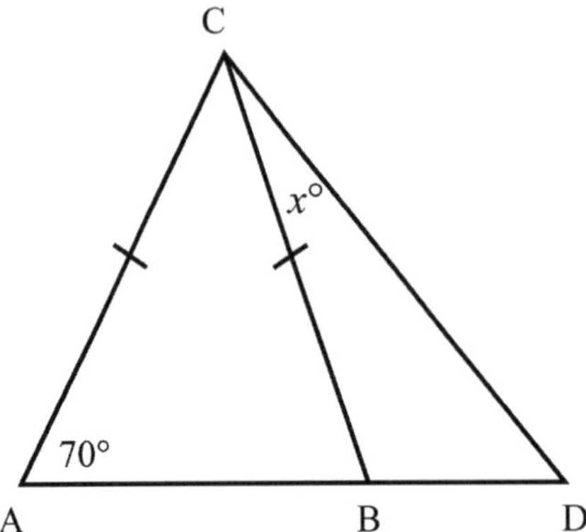

Note:Figure not drawn to scale.

Ⓐ 15°

Ⓑ 20°

Ⓒ 25°

Ⓓ 30°

8.

In a recent survey of two popular best-selling books, two-fifths of the 2,200 polled said they did not enjoy the second book, but did enjoy the first book. Of those, 40% were adults over 18. If three-eighths of those surveyed were adults over 18, how many adults over 18 did NOT report that they enjoyed the first book but not the second book? Assume that all people polled like only one best-selling book among two popular best-selling books.

Ⓐ 187

Ⓑ 352

Ⓒ 473

Ⓓ 626

9.

The diameter of a circle is increased by 80%. By what percent is the area increased?

Ⓐ 80%

Ⓑ 224%

Ⓒ 280%

Ⓓ 325%

10.

Shauna is throwing a bachelorette party for her best friend with x guests total. All of the guests plan to split the cost of renting a limo for y dollars. The day before, z guests cancel. Which of the following represents the percent increase in the amount each guest must pay towards the limo rental?

Ⓐ $\frac{y}{x}$

Ⓑ $\frac{y(x-z)}{x(y-x)}$

Ⓒ $\frac{100yz}{y(x-z)}$

Ⓓ $\frac{100x}{y(z-x)}$

11.

In the f igure, t he radius of circle B is three−fourths the radius of circle A. The distance from center to the of circle B is 3, and the distance from center B to the circumference of circle A is 2. What is the area of the smaller circle?

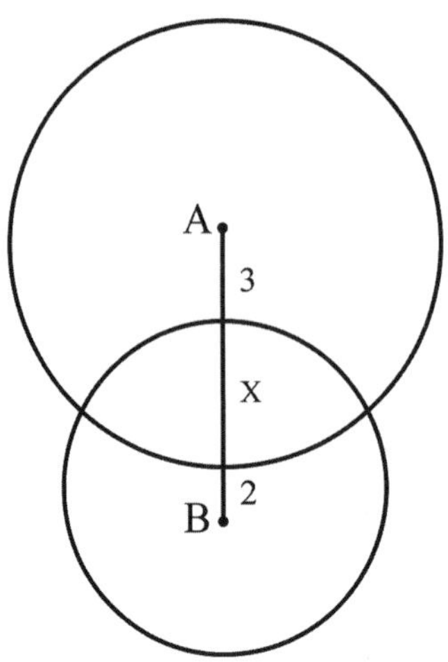

Ⓐ π

Ⓑ 3π

Ⓒ 6π

Ⓓ 9π

12.

The average of several exam scores is 80. One make−up exam was given. Included with the other scores, the new average was 84. If the score on the make−up exam was 92, how many total exams were given?

Ⓐ 2

Ⓑ 3

Ⓒ 4

Ⓓ 5

13.

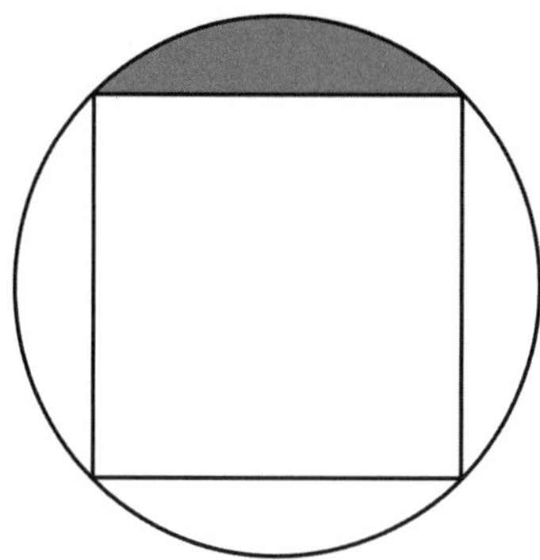

In the figure above, a square is inscribed in a circle. If the area of the square is 36, what is the perimeter of the shaded region?

Ⓐ $6+\frac{3\sqrt{2}}{2}\pi$

Ⓑ $6+3\pi$

Ⓒ $6+3\sqrt{2}\pi$

Ⓓ $6+6\sqrt{2}\pi$

14.

On a coordinate plane, (a, b) and $(a+5, b+c)$, and $(13, 10)$ are three points on line l. If the x-intercept of line l is -7, what is the value of c?

Ⓐ 1.5

Ⓑ 2.0

Ⓒ 2.5

Ⓓ 3.0

15.

If Eric was 22 years old x years ago and Shelley will be 24 years old in y years, what was the average of their ages 4 years ago?

Ⓐ $\dfrac{x+y}{2}$

Ⓑ $\dfrac{y+38}{3}$

Ⓒ $\dfrac{x-y+38}{4}$

Ⓓ $\dfrac{x-y+38}{2}$

16.

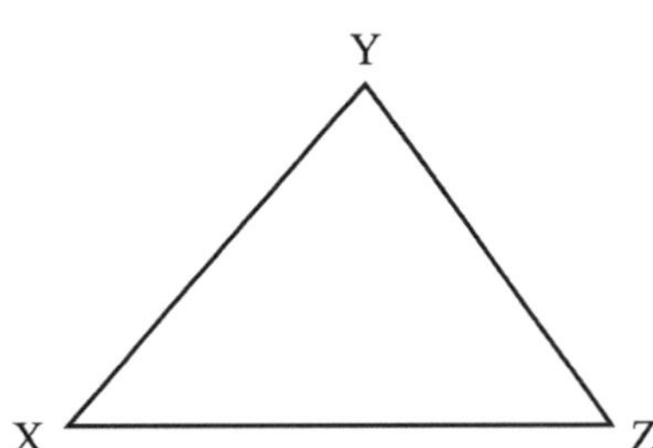

Point A lies between X and Y. Point B lies on YZ and Point C lies on XZ. BZ is congruent to CZ, and $\angle XYZ = 90°$. $XZ - CZ = XA$. What is the value of $\angle ACB$?

Ⓐ $45°$

Ⓑ $60°$

Ⓒ $75°$

Ⓓ $90°$

17.

The points A, B, and C lie on the same circle. If $A=(3,13)$, $B=(15,5)$ and $C=(20,30)$, which of the following could be the center of the circle?

Ⓐ $(15,18)$

Ⓑ $(16,17)$

Ⓒ $(15,13)$

Ⓓ $(17,15)$

18.

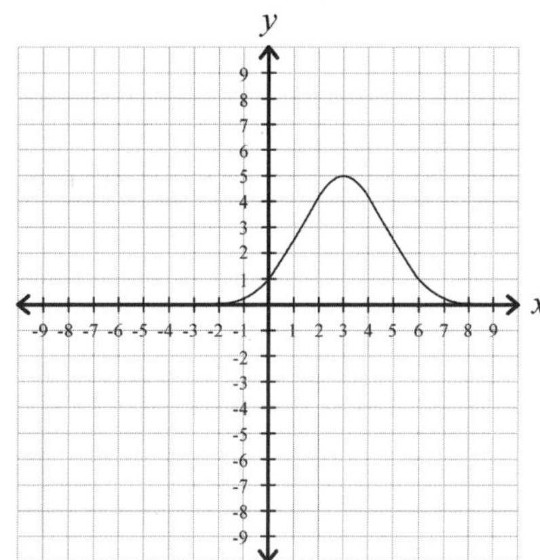

The graph of $y=h(x)$ is shown above. It is true that $f(x)=\dfrac{1}{2}h(-\dfrac{x}{2})$. Which of the following represents the graph of $y=f(x)$?

Ⓐ

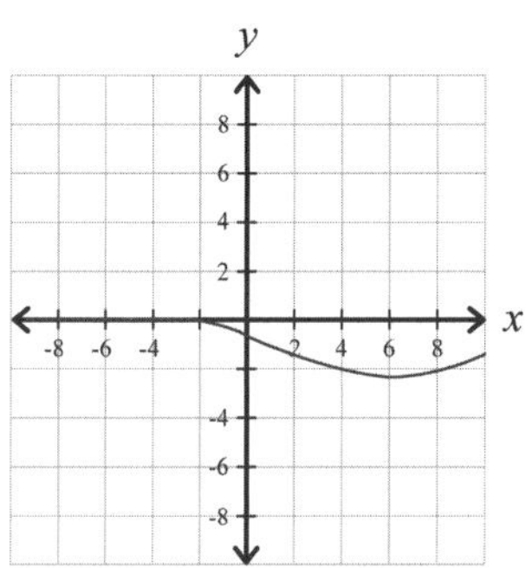

Ⓑ

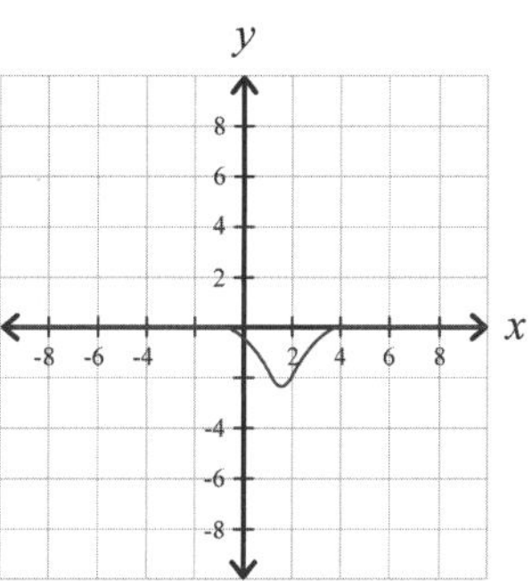

Ⓒ

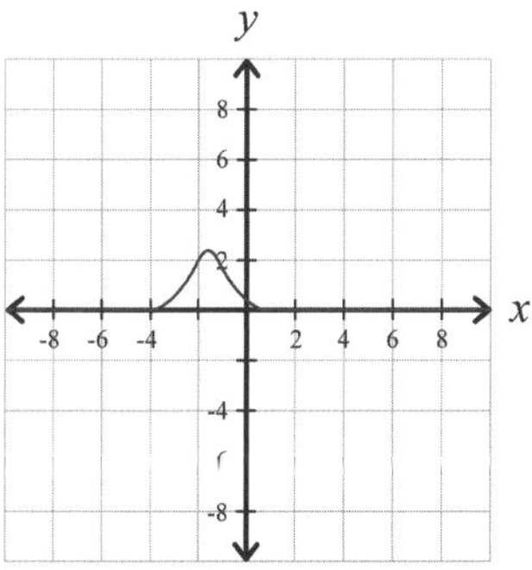

Ⓓ

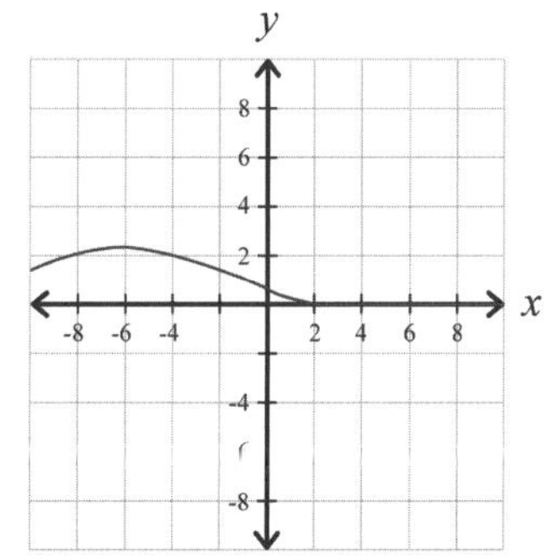

19.

In the xy-coordinate plane, a circle with center $(-4,0)$ is tangent to the line $y=-x$. What is the circumference of the circle?

Ⓐ 2π

Ⓑ $2\pi\sqrt{2}$

Ⓒ 4π

Ⓓ $4\pi\sqrt{2}$

20.

Consider two parallel lines F and G shown below. Line F forms an angle of 60° with the x-axis. Line G has a y-intercept of -3. What is the shortest distance between the two parallel lines?

Ⓐ 3

Ⓑ $\sqrt{3}$

Ⓒ $\dfrac{3\sqrt{3}}{2}$

Ⓓ $\dfrac{3}{2}$

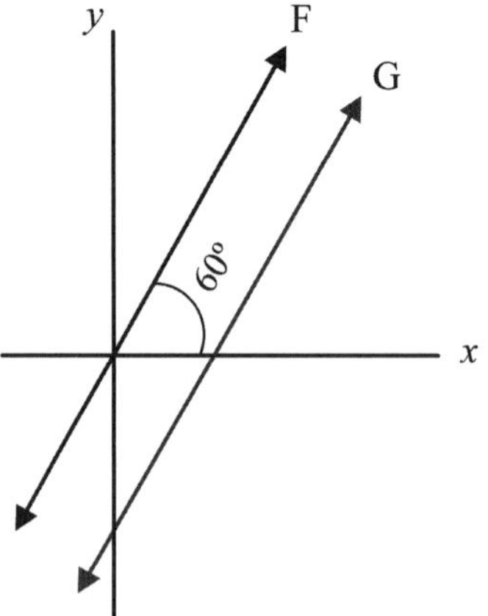

21.

The value of a is chosen such that the product of $2+3i$ and $3+ai$, where $i=\sqrt{-1}$, is a real number (i.e. not complex). What will this product be?

Ⓐ $\frac{27}{2}$

Ⓑ 3

Ⓒ $\frac{39}{2}$

Ⓓ $-\frac{9}{2}$

22.

Which of the following is the value of $\sin(\frac{5\pi}{6})$?

Ⓐ $-\frac{\sqrt{3}}{2}$

Ⓑ $-\frac{\sqrt{2}}{2}$

Ⓒ $\frac{1}{2}$

Ⓓ $\frac{\sqrt{2}}{2}$

Final Practice Test 2

(Easy/Medium)

Math

22 Questions

DIRECTIONS

The questions in this section address a number of important math skills.
Use of a calculator is permitted for all questions.

NOTES

Unless otherwise indicated:

- All variables and expressions represent real numbers.
- Figures provided are drawn to scale.
- All figures lie in a plane.
- The domain of a given function fis the set of all real numbers x for which f(x) is a real number.

REFERENCE

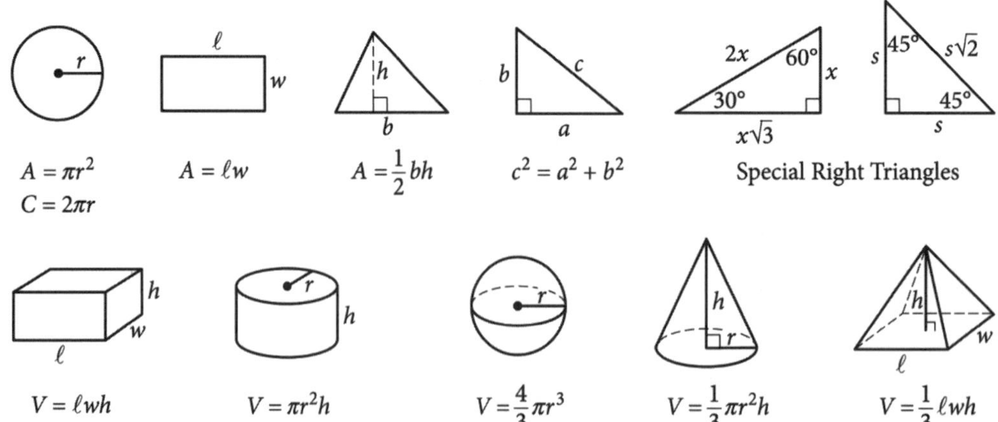

The number of degrees of arc in a circle is 360.
The number of radians of arc in a circle is 2π.
The sum of the measures in degrees of the angles of a triangle is 180.

For multiple-choice questions, solve each problem, choose the correct answer from the choices provided, and then circle your answer in this book.

Circle only one answer for each question. If you change your mind, completely erase the circle. You will not get credit for questions with more than one answer circled, or for questions with no answers circled.

For student-produced response questions, solve each problem and write your answer next to or under the question in the test book as described below.

- Once you've written your answer, circle it clearly. You will not receive credit for anything written outside the circle, or for any questions with more than one circled answer.

- If you find more than one correct answer, write and circle only one answer.

- Your answer can be up to 5 characters for a positive answer and up to 6 characters (including the negative sign) for a negative answer, but no more.

- If your answer is a fraction that is too long (over 5 characters for positive, 6 characters for negative), write the decimal equivalent.

- If your answer is a decimal that is too long (over 5 characters for positive, 6 characters for negative), truncate it or round at the fourth digit.

- If your answer is a mixed number (such as $3\frac{1}{2}$), write it as an improper fraction ($7/2$) or its decimal equivalent (3.5).

- Don't include symbols such as a percent sign, comma, or dollar sign in your circled answer.

1.

If David has twice as many nickels as Tom, and Tom has 15 more nickels than John, what is the value in dollars of David's nickels if John has 6 nickels?

Ⓐ $1.40

Ⓑ $2.10

Ⓒ $21

Ⓓ $42

2.

There are 5 pencil-cases on the desk. Each pencil-case contains at least 10 pencils, but not more than 14 pencils. Which of the following could be the total number of pencils in all 5 cases?

Ⓐ 35

Ⓑ 45

Ⓒ 65

Ⓓ 75

3.

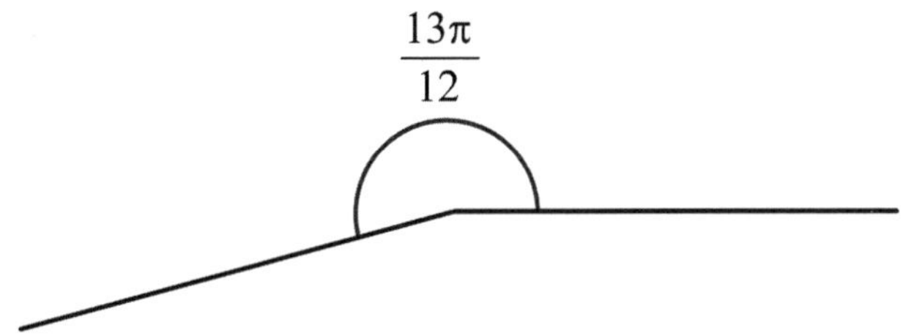

Which of the following is the measure of the above angle in degrees?

Ⓐ $185°$

Ⓑ $195°$

Ⓒ $210°$

Ⓓ $225°$

4.

A cylindrical birthday cake with a height of 4 inches is cut into two pieces such that each piece is of a different size. If the ratio of the volume of the larger slice to the volume of the smaller slice is 5 to 3, what is the degree measure of the cut made into the cake?

Ⓐ $115°$

Ⓑ $120°$

Ⓒ $135°$

Ⓓ $145°$

5.

$$\sqrt{72} + \sqrt{72} = m\sqrt{n}$$

In the equation above, m and n are integers such that $m > n$. Which of the following is the value of m ?

Ⓐ 6

Ⓑ 12

Ⓒ 16

Ⓓ 24

6.

In the xy-plane, line l passes through the point $(0, 1)$ and is parallel to the line with equation $2x + 3y = 6$. If the equation of line l is $y = ax + b$, what is the value of $a + b$?

7.

The line represented by the equation $y = \frac{1}{12} - x$ is graphed in the xy-plane. Which of the following statements correctly describes the graph of the line?

Ⓐ The line is perpendicular to the graph $x+y=1$.

Ⓑ The line has a negative slope and a positive y-intercept.

Ⓒ The line has a positive slope and a negative y-intercept.

Ⓓ The x-intercept is equal to the negative of the y-intercept.

8.

$$3x^2+kx=-3$$

What value of k will result in exactly one solution to the equation?

Ⓐ 9

Ⓑ -6

Ⓒ 10

Ⓓ -10

9.

Let $g(x)=8x-5$. Which of the following is equivalent to $g(g(x))$?

Ⓐ $64x-10$

Ⓑ $64x-45$

Ⓒ $64\ x^2\quad+25$

Ⓓ $64\ x^2\quad-80+25$

10.

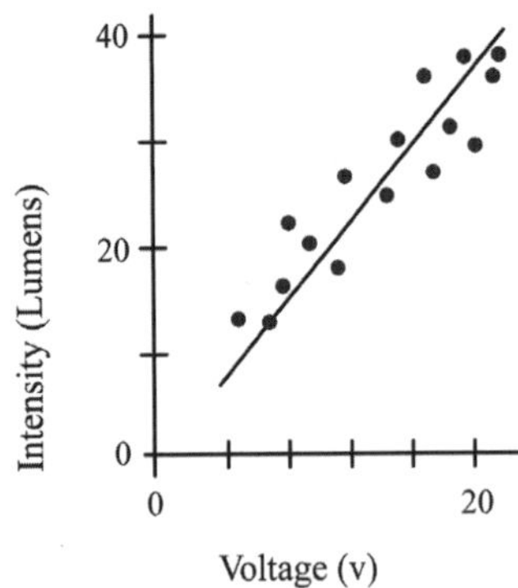

The figure shows the line of best fit through a set of experimental data. If the equation of the line of best fit is:

$$\text{Intensity} = 1.78\times\ (\text{Voltage}) + 1.5$$

Which of the following input voltages would most likely result in an Intensity of 30 Lumens according to the line of best fit?

Ⓐ 15.3 Ⓑ 16.0 Ⓒ 20.0 Ⓓ 54.9

11.

A pool that holds 35,000 cubic feet of water is being filled by a pump at a rate of 200 cubic feet per minute. At the same time, water is draining out through an open valve accidentally left open. If the pool is full in 200 minutes, how many cubic feet of water are draining out per minute?

Ⓐ 5

Ⓑ 15

Ⓒ 25

Ⓓ 30

12.

In the given figure, the measure of angle OAC is 60 degrees, and the center of the circle is O. If the circle has a radius of 6, what is the length of segment DB?

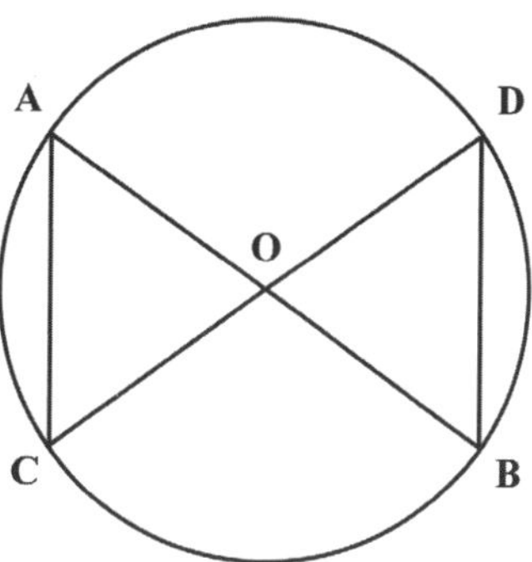

Note: Figure not drawn to scale.

Ⓐ 3

Ⓑ $3\sqrt{2}$

Ⓒ 6

Ⓓ $6\sqrt{2}$

13.

If $3x-y=12$, what is the value of $\frac{8^x}{2^y}$

Ⓐ 2^{12}

Ⓑ 4^4

Ⓒ 8^2

Ⓓ The value cannot be determined from the information given.

14.

For each of the largest 21 airline companies in Europe, the average delay of flights was calculated and shown in the dot plot above. If we remove the two airlines with the highest flight delays from the dot plot, which of the following will result?

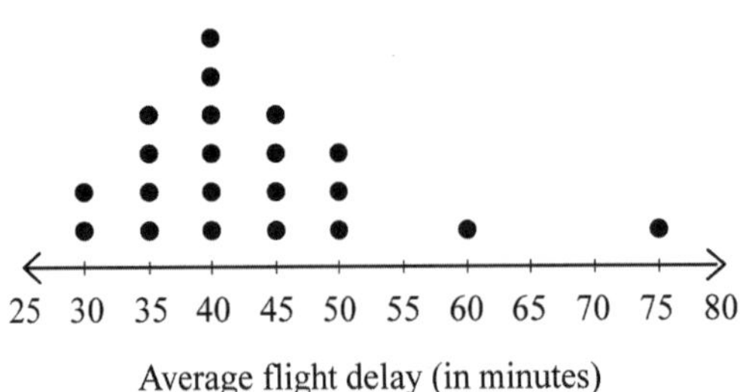

Ⓐ The mean will decrease only.

Ⓑ The mean and range will decrease only.

Ⓒ The mean and median will decrease only.

Ⓓ The mean, median, and range will decrease.

15.

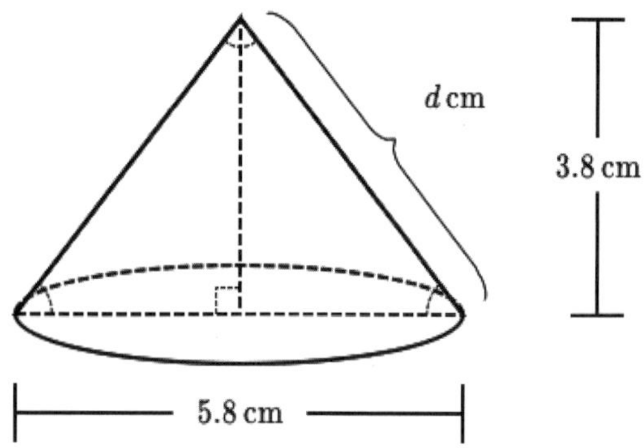

The approximate height and base diameter of a conical metal valve plug are shown in the diagram above in centimeters (cm). What is the distance, d, in centimeters, from the vertex to the outer edge of the circular base? (Round your answer to the nearest tenth of a centimeter.)

16.

A company uses an additive manufacturing ("3D−Printing") process to create square pyramids as shown in the illustration. If the 3D−printer can print $0.025\ cm^3$ per second, how many minutes will it take to print one pyramid?

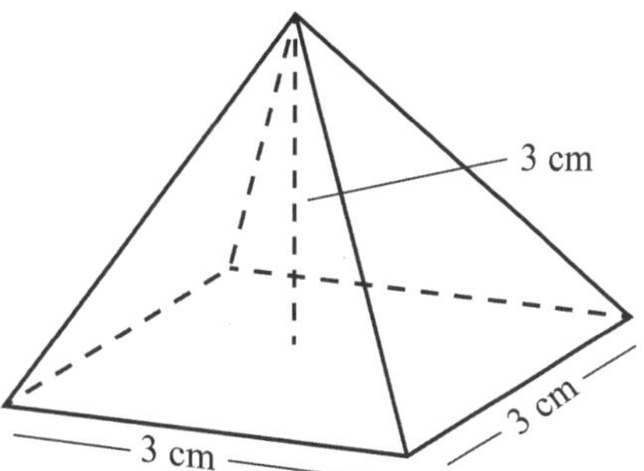

Ⓐ 14

Ⓑ 9

Ⓒ 6

Ⓓ 18

17.

The functions $g(x)=2(x-5)(x-3)$ and $h(x)=2(x+5)(x+3)$ are graphed in the xy-plane. Which of the following is a true statement?

Ⓐ The functions have the same y-intercept.

Ⓑ The functions have the same x-intercepts.

Ⓒ The functions have the same axis of symmetry.

Ⓓ The functions have the same vertex.

18.

On a certain street there are 7 houses. The value of each of these houses is provided in the table below. An 8th house is being added on the same street and will have a value in excess of $255,000. What is the lowest value that this new home can have such that the mean of all 8 house values will be greater than or equal to the median of the 8 house values?

	Value
House 1	$180,000
House 2	$200,000
House 3	$225,000
House 4	$250,000
House 5	$252,000
House 6	$255,000
House 7	$256,000
House 8	?

Ⓐ $390,000

Ⓑ $502,000

Ⓒ $415,000

Ⓓ $276,000

19.

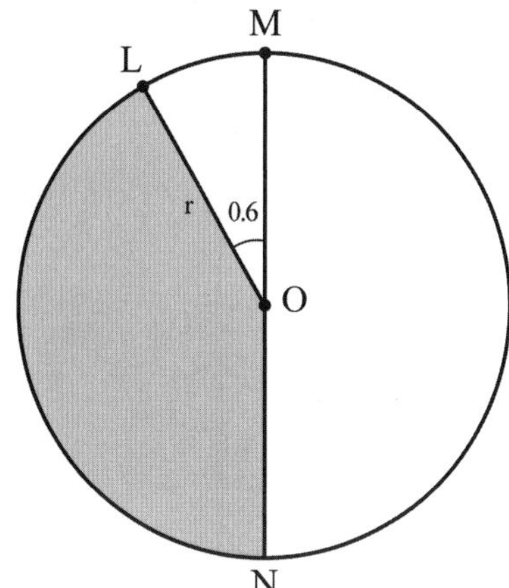

The circle shown at the above has its center at point O. Line segment is a diameter. The measure of acute angle LOM is 0.6 radians. The shaded sector of the circle formed by the obtuse angle LON has area 6. What is the radius, r, of the circle?

Ⓐ $\dfrac{3}{\pi - 0.6}$

Ⓑ $\dfrac{12}{\pi - 0.6}$

Ⓒ $\sqrt{\dfrac{3}{\pi - 0.6}}$

Ⓓ $\sqrt{\dfrac{12}{\pi - 0.6}}$

20.

$$v(t) = 0.41 \times 0.8^{\frac{2t}{5}}$$

A researcher studied the ability of a gelatin sample to retain an electrical charge. The function above approximates the voltage, in volts, of the sample t minutes after it has been charged. Which of the following statements best describes the relationship between the voltage and the number of minutes?

Ⓐ It is linear because the voltage increases by 0.41 volts every 2.5 minutes.

Ⓑ It is linear because the voltage decreases by 0.8 volts every 0.4 minutes.

Ⓒ It is exponential because the voltage decreases by 20% every 2.5 minutes.

Ⓓ It is exponential because the voltage increases by 20% every 0.4 minutes.

21.

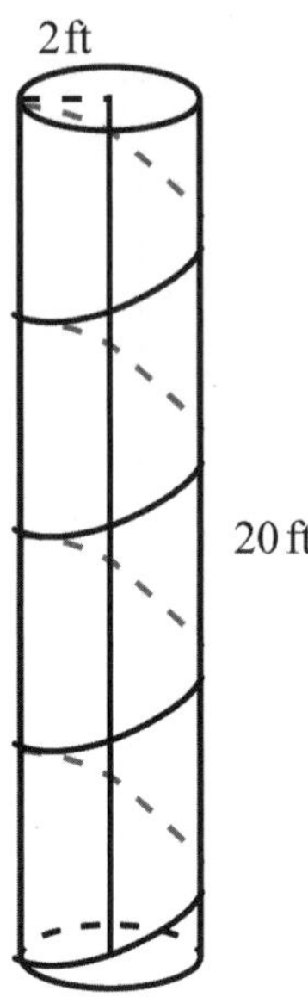

A spiral staircase makes 4 full turns as it climbs 20 feet (ft), as represented in the figure above. The outer railing is a constant distance of 2 ft from the midline of the staircase. About what is the length of the railing?

Ⓐ $32ft$ Ⓑ $50ft$ Ⓒ $54ft$ Ⓓ $80ft$

22.

A circle in the xy−plane has its center on the line $x=3$. If the point $(4,5)$ lies on the circle and the radius is, which of the following could be the center of the circle?

Ⓐ $(3,3)$

Ⓑ $(3,4)$

Ⓒ $(3,5)$

Ⓓ $(3,7)$

Final Practice Test 2

(Advanced/Difficult)

Math

22 Questions

DIRECTIONS

The questions in this section address a number of important math skills.
Use of a calculator is permitted for all questions.

NOTES

Unless otherwise indicated:

- All variables and expressions represent real numbers.
- Figures provided are drawn to scale.
- All figures lie in a plane.
- The domain of a given function fis the set of all real numbers x for which f(x) is a real number.

REFERENCE

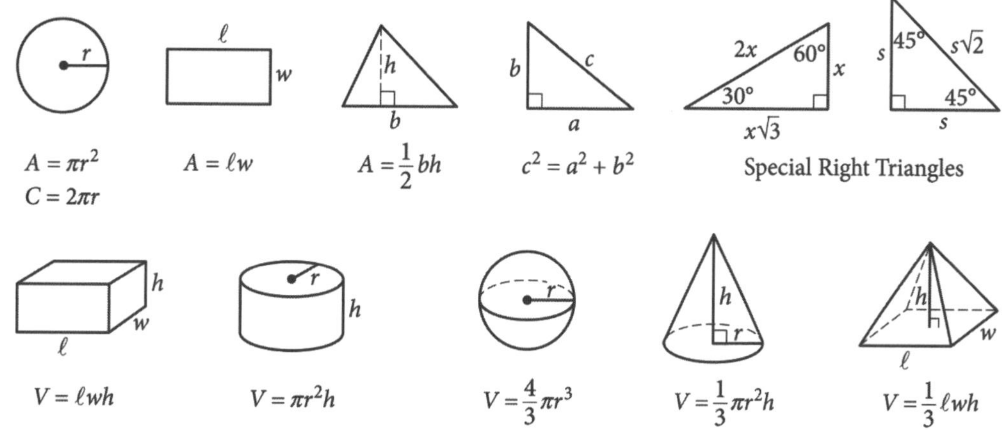

The number of degrees of arc in a circle is 360.
The number of radians of arc in a circle is 2π.
The sum of the measures in degrees of the angles of a triangle is 180.

For multiple-choice questions, solve each problem, choose the correct answer from the choices provided, and then circle your answer in this book.
Circle only one answer for each question. If you change your mind, completely erase the circle. You will not get credit for questions with more than one answer circled, or for questions with no answers circled.

For student-produced response questions, solve each problem and write your answer next to or under the question in the test book as described below.

- Once you've written your answer, circle it clearly. You will not receive credit for anything written outside the circle, or for any questions with more than one circled answer.

- If you find more than one correct answer, write and circle only one answer.

- Your answer can be up to 5 characters for a positive answer and up to 6 characters (including the negative sign) for a negative answer, but no more.

- If your answer is a fraction that is too long (over 5 characters for positive, 6 characters for negative), write the decimal equivalent.

- If your answer is a decimal that is too long (over 5 characters for positive, 6 characters for negative), truncate it or round at the fourth digit.

- If your answer is a mixed number (such as $3\frac{1}{2}$), write it as an improper fraction ($7/2$) or its decimal equivalent (3.5).

- Don't include symbols such as a percent sign, comma, or dollar sign in your circled answer.

1.

At a certain lab, the current ratio of scientists to engineers is $5:1$. The lab is considering a plan where 75 new team members would be hired in the ratio of two engineers per scientist, which would change the ratio of scientists to engineers to approximately $2:3$. Approximately how many scientists currently work at the lab?

Ⓐ 5

Ⓑ 10

Ⓒ 15

Ⓓ 20

2.

Officials project that between 2010 and 2050, the Sub–Saharan African population will drastically change. The model below gives the projection of the population, P, in thousands, with respect to time, t(provided that 2010 is when $t=0$).

$$P=175+\frac{11}{2}t$$

What does the 175 mean in the equation?

Ⓐ In 2010, the population of Sub–Saharan African was 175 thousand.

Ⓑ In 2050, the population of Sub–Saharan African will be 175 thousand

Ⓒ Between 2010 and 2050, the population of Sub–Saharan African will increase by 175 thousand.

Ⓓ Between 2010 and 2050, the population of Sub–Saharan African will decrease by 175 thousand.

3.

Two lines graphed in the xy-plane have the equations $2x+5y=20$ and $y=kx-3a$, where k is a constant. For what value of ka will the two lines be the same lines?

Ⓐ $-\frac{2}{3}$

Ⓑ $\frac{2}{3}$

Ⓒ $-\frac{8}{15}$

Ⓓ $\frac{8}{15}$

4.

Populations of Groups A and B can be modeled exponentially. The increasing population of Group A is described by:

$$P_A = 100e^{0.02t}$$

The decreasing population of Group B is described by:

$$P_B = 2000e^{-0.01t}$$

In both equations t represents time in minutes. At what time (in minutes) will the Group A population be twice that of Group B?

Ⓐ 77

Ⓑ 100

Ⓒ 123

Ⓓ 146

5.

A bag contains tomatoes that are either green or red.

The ratio of green tomatoes to red tomatoes in the bag is 4 to 3. When five green tomatoes and five red tomatoes are removed, the ratio becomes 3 to 2. How many red tomatoes were originally in the bag?

Ⓐ 12

Ⓑ 15

Ⓒ 18

Ⓓ 24

6.

$$a, 2a-1, 3a-2, 4a-3, \ldots$$

For a particular number a, the first term in the sequence above is equal to a, and each term there after is 7 greater than the previous term. What is the value of the 16th term in the sequence?

7.

If p is a prime number, how many factors does p^3 have?

Ⓐ One

Ⓑ Two

Ⓒ Three

Ⓓ Four

8.

If a and b are numbers such that $(a-4)(b+6)=0$, then what is the smallest possible value of a^2+b^2 ?

9.

Let $f(x) = ax^2$ and $g(x) = bx^4$ for any value of x. If a and b are positive constants, for how many values of x is $f(x) = g(x)$?

Ⓐ None

Ⓑ One

Ⓒ Two

Ⓓ Three

10.

The container shown below is filled such that the height of water in the container increases at a constant rate. Which graph best describes the rate at which water is being added versus time?

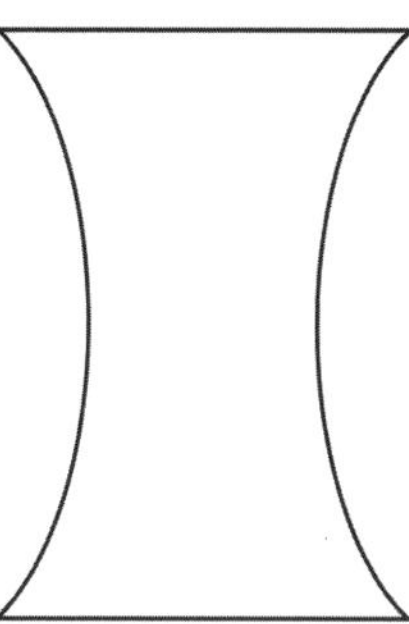

Ⓐ

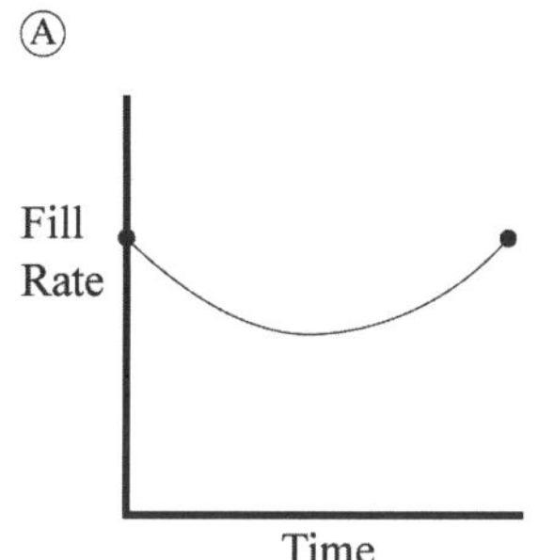

Ⓑ

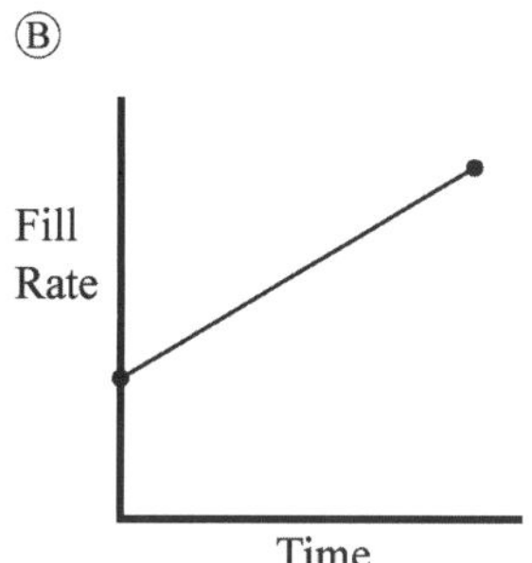

Ⓒ

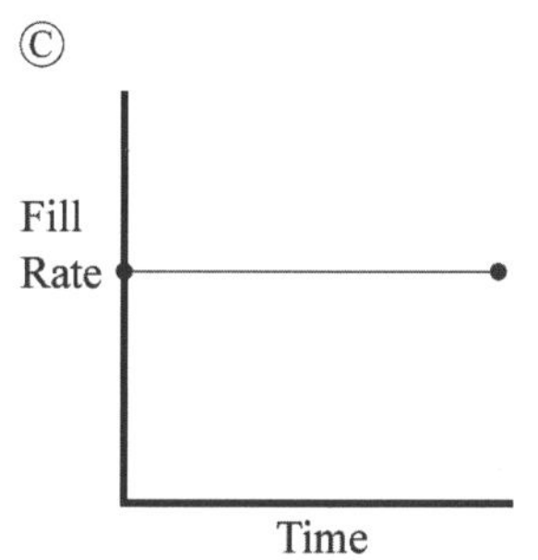

Ⓓ

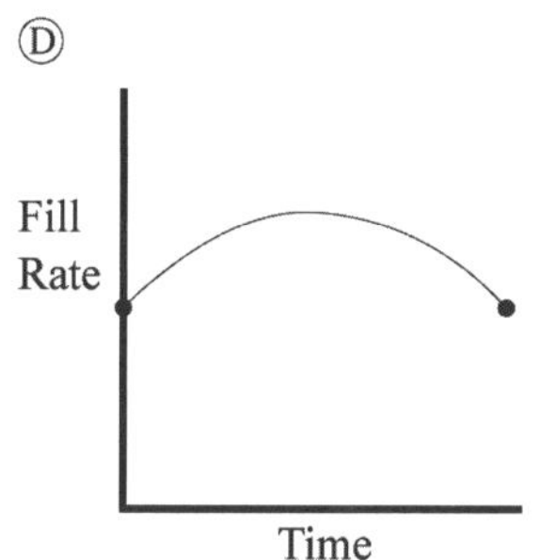

11.

t	0	1	2
$N(t)$	128	16	2

The table above shows some values for the function $N(t)$.

If $N(t)=k\cdot 2^{-at}$ for positive constants k and a, what is the value of a ?

Ⓐ -2

Ⓑ $\frac{1}{3}$

Ⓒ 2

Ⓓ 3

12.

The value of y increased by 12 is directly proportional to the value of x decreased by 6. If $y=2$ when $x=8$, what is the value of x when $y=16$?

Ⓐ 8

Ⓑ 10

Ⓒ 16

Ⓓ 20

13.

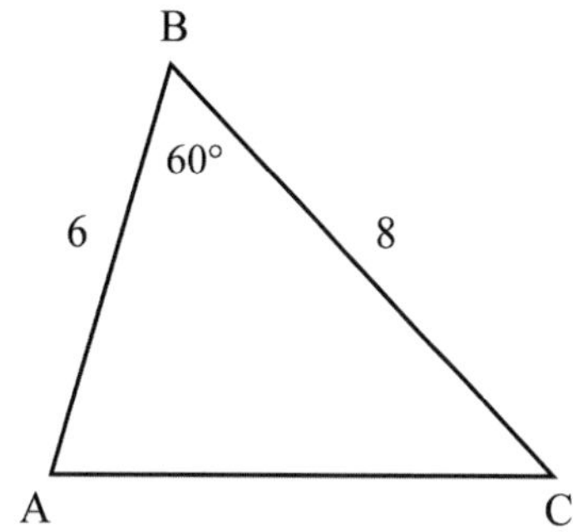

In the figure above, $AB=6$ and $BC=8$. What is the area of triangle ABC ?

Ⓐ $12\sqrt{2}$

Ⓑ $12\sqrt{3}$

Ⓒ $24\sqrt{2}$

Ⓓ $24\sqrt{3}$

14.

Line l goes through points P and Q, whose coordinates are $(0,1)$ and $(b,0)$, respectively. For which of the following values of b is the slope of line l greater than $-\frac{1}{2}$

Ⓐ 1

Ⓑ $\frac{3}{2}$

Ⓒ $\frac{5}{3}$

Ⓓ $\frac{5}{2}$

15.

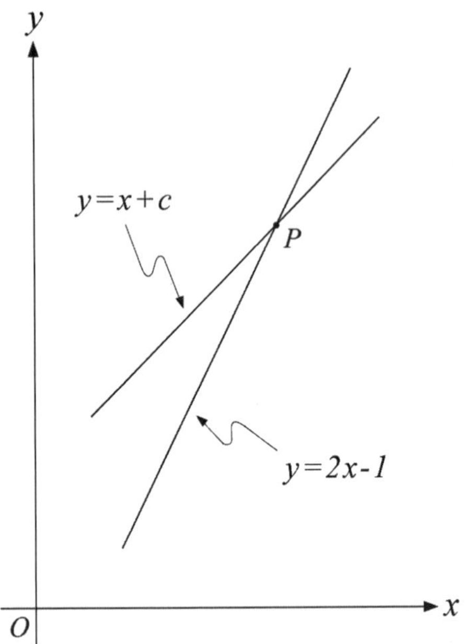

Note: Figure not drawn to scale.

In the xy-plane, the lines $y=2x-1$ and $y=x+c$ intersect at point P, where c is a positive number. Portions of these lines are shown in the figure above. If the value of c is between 1 and 2, what is one possible value of the x-coordinate of P?

16.

VitaDrink contains 30 percent concentrated nutrients by volume. EnergyPlus contains 40 percent concentrated nutrients by volume. Which of the following represents the percentage of concentrated nutrients by volume in a mixture of v gallons of VitaDrink, e gallons of EnergyPlus, and w gallons of water?

Ⓐ $\dfrac{v+e}{v+e+w}$

Ⓑ $\dfrac{0.3v+0.4e}{v+e+w}$

Ⓒ $\dfrac{3v+4e}{v+e+w}$

Ⓓ $\dfrac{30v+40e}{v+e+w}$

17.

The average (arithmetic mean) of a particular set of seven numbers is 12. When one of the numbers is replaced by the number 6, the average of the set increases to 15. What is the number that was replaced?

Ⓐ -20

Ⓑ -15

Ⓒ -12

Ⓓ 0

18.

Let a, b and c be positive integers. If the average (arithmetic mean) of a, b and c is 100, which of the following is NOT a possible value of any of the integers?

Ⓐ 1

Ⓑ 100

Ⓒ 297

Ⓓ 299

19.

M is a set consisting of a finite number of consecutive integers. If the median of the numbers in set M is equal to one of the numbers in set M , which of the following must be true?

I. The average (arithmetic mean) of the numbers in set M equals the median.

II. The number of numbers in set M is odd.

III. The sum of the smallest number and the largest number in set M is even.

Ⓐ I only

Ⓑ II only

Ⓒ II and III only

Ⓓ I, II and III

20.

$$4|6+2s|-27 \leq -3$$

Which of the following best describes the solutions to the inequality shown above?

Ⓐ $-24 \leq s \leq 0$

Ⓑ $-6 \leq s \leq 0$

Ⓒ $s \leq 0$ or $s \geq 3$

Ⓓ No solution

21.

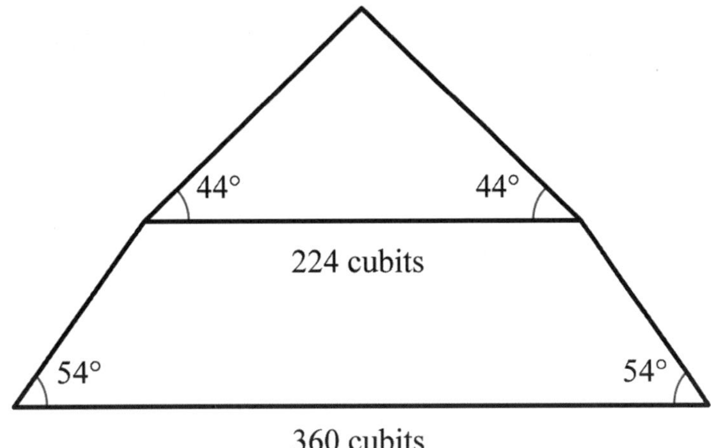

According to some reconstructions, the Sneferu pyramid, also called the bent pyramid, had the measurements shown above. Segments that appear parallel are parallel. Based on those measurements, what was the height of the pyramid to the nearest cubit?

Note:

$\cos(44°) \approx 0.72$, $\tan(44°) \approx 0.97$, $\cos(54°) \approx 0.59$ and $\tan(54°) \approx 1.38$.

22.

Two cars are racing at a constant speed around a circular racetrack. Car A requires 15 seconds to travel once around the racetrack, and car B requires 25 seconds to travel once around the racetrack. If car A passes car B, how many seconds will elapse before car A once again passes car B ? Round to the nearest tenths.

How to use a calculator on The Digital SAT Math (Using by Ti−84 Plus/Ti−Nspire)

Basic Operation

1. $-(-312-114)=$

2. $\dfrac{64}{9\pi}=$

3. $\dfrac{13\times(-41)}{12\pi-43}=$

4. $2+4+6+8+10+\cdots+100=$

Exponential Operation

5. $3^4+3^{-3}+(-3^2)+(-3)^{-2}=$

6. $10\times(1-0.2)^4=$

7. $\sqrt[3]{50^2}=$

8. $1.3^{-\frac{11}{12}}=$

Sketching the Graph of Function

9. $y = -3x + 4$

10. $y = x^2$

11. Find the values of x in the following equation.

$$x^2 = -3x + 4$$

12. $y = 2x^2 - 3x - 5$

i) Vertex :

ii) y intercept :

iii) x intercept :

13. $y = 2 \times 3^x$

14. Find the values of x in the following equation.

$$2 \times 3^x = 400 + \pi$$

Finding the solution of Equation by Graphing

15.

$$20+\frac{800}{x}=\frac{800}{x-2}$$

Line Regression

16.

Townsend Realty Group Investments		
Property address	Purchase price (dollars)	Monthly rental price (dollars)
Clearwater Lane	128,000	950
Driftwood Drive	176,000	1,310
Edgemont Street	70,000	515
Glenview Stereet	140,000	1,040
Hamilton Circle	450,000	3,365

The Townsend Realty Group Invested in the five different proporties listed in the table above. The table shows the amount, in dollars, the company paid for each property and the corresponding monthly rental price, in dollars, the company charges for the property at each of the five locations. The relationship between the monthly rental price, r, in thousands of dollars, can be represented by a linear function. Which of the following functions represents the relationship?

Ⓐ $r(p)=2.5p-870$

Ⓑ $r(p)=5p+165$

Ⓒ $r(p)=6.5+440$

Ⓓ $r(p)=7.5p-10$

ANSWERS KEY & EXPLANATIONS

ANSWERS KEY AND EXPLANATIONS

Digital SAT MATH 200 ESSENTIAL PRACTICE QUESTIONS FOR FULL MARKS

QUESTION 1

Choice C

$4+5m=5m+1$

$5m-5m=4-1$

Since $0\times m=3$

$\therefore$ This linear equation has no solution.

QUESTION 2

Choice C

$-4+bx=2x+3x+3$

$(b-5)x=7$

For this linear equation has infinitely many solutions

$b-5=0$

$\therefore b=5$

QUESTION 3

Choice B

On $y=\frac{2}{1}x+3$

2 is the rate of chang in y and 1 is the rate of change in x.

QUESTION 4

Choice D

On given the equation, $y=-5$

this equation is the same formation as $y=0\times x-5$.

so when $x=0$, the initial value of y is -5.

QUESTION 5

Choice D

The y−intercept is a value of y when $x=0$.

QUESTION 6

Choice B

On the equation $2x+3y=12$,

2 is the rate of change in x and 3 is the rate of change in y.

QUESTION 7

Choice A

The highest degree of $y=4x^2+2x-3$ is 2.

So this equation is a quadratic, not a linear.

QUESTION 8

Choice C

$$\text{Slope} = \frac{6}{3} = \frac{t}{s}$$

$6s = 3t$

$2s = t$

$\therefore t : s = 2 : 1$

QUESTION 9

Choice B

The rate of change of T is $+10$.

This means the interior temperature of the steak increase by 10 per every minute.

QUESTION 10

Choice A

On $P=\frac{11}{2}t+175$

175 means the initial value of population of P when $t=0$ in 2010.

QUESTION 11

Choice B

The Number of people in swimming pool is $g(t)$.

For finding the value of c when $g(t)=350$ and $t=90$

$$350 = \frac{5(7\times 90 - 12c)}{2} - 25$$

We can solve this equation and find the answer.

$\therefore c = 40$

QUESTION 12

Choice D

Since the caricature artist charges \$35 per hours and A means the number of hours to hire this caricature artist, $35A$ is the amount of total costs of the caricature artist.

QUESTION 13

Answer: 8 and 4

Let's assume two numbers are x and y where $x > y$.

Then $x+y=12$ and $x-y=4$.

If we add up the sides of the two equations,

$\therefore x=8$ and $y=4$.

QUESTION 14

Answer: 39

Let the number of adult tickets : x

and the number of student tickets : $60-x$

Since adult tickets cost \$15 each and student tickets cost \$8

$15x+8(60-x)=750$ (total cost)

$7x = 750-480$

$\therefore x=38.57 \fallingdotseq 39$ as the nearest whole number.

QUESTION 15

Answer: 23

Let the number of brand B laptops : x

and the number of brand A laptops : $2x$

Since brand A laptops sell for \$1000 each and brand B laptops sell for \$800

$800 \times (2x) + 1000 \times x = 60{,}000$ (total cost)

$2600x = 6000$

$\therefore x = 23.08 \fallingdotseq 23$ as the nearest whole number.

QUESTION 16

Answer: $A = 300$ and $B = 200$

Let the number of product A : x

and the number of product B : y

According to the conditions of the question,

$x + y = 500$

$2x + 3y = 1200 \quad \therefore x = 300$ and $y = 200$

QUESTION 17

Choice B

Let the number of times that Joanne volunteer : $x + 4$

and the number of times that Richard volunteer : x

According to the condition the sum of two number is not more than a combined total of 16 hours

$(x+4) + x \leq 16$

$\therefore 2x + 4 \leq 16$

QUESTION 18

Answer: 4

Let the number of times that Maria rent a boat : t

According to the conditions of the question,

$60t + 10 \leq 280$

$60t \leq 270$

$$t \leq \frac{27}{6} = 4.5$$

Since t is a positive integer $\therefore t = 4$

QUESTION 19

Choice B

From given that $D = 60 - \frac{3}{4}P$, $S = \frac{1}{4}P$ and $D = S$

$60 - \frac{3}{4}P = \frac{1}{4}P$

$\therefore P = 60$

QUESTION 20

Choice D

[1.5 hours (time coming to the park per day) + 7 hours (volunteer time per day) + 1.5 hours (time going back per day)]$\times d$ days $\geq$ 35 hours

QUESTION 21

Choice B

Rate of change in (−)4 $^{\circ}F$ every 1000 feet from ground height : slope $= \frac{-4}{1000}$

Initial value of temperature at ground : 70

$\therefore y$(temperature) $= -\frac{4}{1000}x$ (height from ground) $+70$

when $y = -58$

$-58 = -\frac{4}{1000}x + 70$

$\therefore 70 = \frac{4}{1000}x - 58$

QUESTION 22

Choice C

The line B $x = -4.5$ is a vertical line to x−axis.

So the line that is perpendicular to the line B is horizontal with slope, 0.

And this line passes through the point $(-1, 3.5)$.

$\therefore y - 3.5 = 0 \times (x - (-1))$

$\therefore y = 3.5$

QUESTION 23

Choice A

From the slope of 4 and x−intercept at $(-2,0)$ form of the line equation,

$y = 4(x-(-2))$

$\therefore y = 4x+8$

QUESTION 24

Choice B

From the x−intercept at $(-15,0)$ and y−intercept at $(0,-9)$ form of the line equation,

$$\frac{x}{(-15)}+\frac{y}{(-9)} = 1$$

$$\therefore \frac{x}{-15}+\frac{y}{-9} =1$$

QUESTION 25

Answer: 5

The slope of the line, $3y = 2x + 8$ is $\frac{2}{3}$.

And the slope between two point $(a,2)$ and $(3,a)$ is $\frac{a-2}{3-a}$.

Since these two lines are perpendicular each other, so the product between slopes is -1.

$$\therefore \frac{2}{3} \times \frac{a-2}{3-a} = -1$$

$2a-4 = -9+3a$

$\therefore a=5$

QUESTION 26

Choice C

Given plane is U,t−plane and $U=9.8h$ is a constant line at this U,t−plane.
So $U=9.8h$ is a horizontal line at U,t−plane.

QUESTION 27

Choice C

The slope of $2x+5y=20$ is $-\frac{2}{5}$ and the slope of $y=kx-3$ is k.

Since these two lines are perpendicular each other, so the product between slopes is -1, $-\frac{2}{5}k = -1$

$\therefore k = \frac{5}{2}$

QUESTION 28

Choice B

The slope of $5x+7y=1$ is $-\frac{5}{7}$ and the slope of $ax+by=1$ is $-\frac{a}{b}$.

Since these two lines are perpendicular each other, so the product between slopes is -1,

$(-\frac{5}{7})(-\frac{a}{b}) = -1$

$\therefore 5a = -7b$

$\therefore 5a+7b=0$

In Choice B, the slope of $10x+7y=1$ is $-\frac{7}{10}$ and the slope of $ax+2by=1$ is $-\frac{2b}{a}$.

Since the product of two slopes is $(-\frac{7}{10})\times(-\frac{2b}{a}) = \frac{14b}{10a} = \frac{7b}{5a}$ anf if these two lines are perpendicular, $\frac{7b}{5a} = -1$

$\therefore 5a = -7b$

$\therefore 5a+7b=0$ So satisfied.

QUESTION 29

Answer: $C = 0.25x + 30$

Rate of change in \$0.25 every mile driven : slope = \$0.25

Initial flat fee : \$30

$\therefore C = 0.25x + 30$

QUESTION 30

Answer: $T = 2.5x + 18$

Rate of change in 2.5 $^{\circ}C$ every hour : slope = 2.5

Initial temperature : 18 $^{\circ}C$

$\therefore T = 2.5x + 18$

QUESTION 31

Answer: $C = 0.10x + 20$

Rate of change in \$0.10 every hour : slope = 0.10

Initial costs of a cell phone plan : \$20

$\therefore C = 0.10x + 20$

QUESTION 32

Choice A

Rate of change of 17 *mph* every 2 hour : slope = $\frac{17}{2} = 8.5$

Initial currently speed of a roller coster : 49*mph*

$\therefore f(t) = 8.5t + 49$

QUESTION 33

Answer: 8

Rate of change of 2 feet per day : slope = 2

Initial currently height of a bamboo plant : 11 feet

$\therefore h = 2t + 11$

So when $h = 27$ $\therefore t = 8$

QUESTION 34

Choice B

Rate of change of $-1.84\,^\circ F$ every 1000 feet : slope = $\frac{-1.84}{1000} = -0.00184$

Initial currently boiling point of water at sea level : $211\,^\circ F$

$\therefore B = -0.00184h + 211$

QUESTION 35

Choice B

The purely silver(100%) amount in 92.5% sterling silver x gram : $0.925x$

The purely silver(100%) amount in 88% silver alloy y gram : $0.88y$

The purely silver(100%) amount in two 91% mixtures of two silvers $x+y$ gram : $0.91(x+y)$

$\therefore$ $0.925x + 0.88y = 0.91(x+y)$

$\therefore\ 0.925x + 0.88y = 0.91x + 0.91y$

$\therefore 0.015x - 0.03y = 0$

QUESTION 36

Choice D

Since this problem lets

First sprinting regimen time : s

Second sprinting regimen time : $s+150$

Third sprinting regimen time : $2s$

And final sprinting regimen time : 80

So total number of sprinting regimen times, $t = s + (s+150) + 2s + 80$

$\therefore t = 4s + 230$

QUESTION 37

Answer:

Let the time of product A requires x hours

and the time of product B requires y hours, then

$x \geq 0,\ y \geq 0$ and $4x + 3y \leq 60$ (labor available constraint)

QUESTION 38

Answer:

Let the amount of apple juice: x liters

and the amount of orange juice: y liters, then

$0 \leq x \leq 100,\ 0 \leq y \leq 150$ and $3x + 2y \geq 300$ (sales goal)

QUESTION 39

Answer:

Let the area of corn planting : x acres

and the area of wheat planting : y acres, then

$x \geq 0,\ y \geq 0$ and $x + y \leq 40$ (land available constraint),

$2x + 3y \leq 100$ (labor available constraint)

QUESTION 40

Answer:

Let the number of standard TV : x

and the number of HD TV : y, then

$x \geq 0$, $y \geq 0$ and $x+y \geq 4$ (product goal),

$500x + 800y \leq 3000$ (budgets available constraint)

QUESTION 41

Choice B

Since $y > 2x-1 > 5-1=4$

$\therefore y > 4$

QUESTION 42

Answer: 22

($\$8 \times 10$ hours + $\$10 \times x$ hours)$\times 0.9 \geq 270$

$$80 + 10x \geq \frac{270}{0.9} = 300$$

$10x \geq 220$

$\therefore x \geq 22$

QUESTION 43

Choice C

Let the number of tutoring time : x hours (pays \$12 per hour)

and the number of lifeguard time : y hours, then (pays \$9.5 per hour)

since Jackie wants to earn at least \$220 : $12x+9.5y \geq 220$

and she can work no more 20 hours : $x+y \leq 20$

So answer is choice C.

QUESTION 44

Answer: 14

If we can let the number of resisters that is need per group: R

and the number of capacitors is need per group: C

Large circuit is needed to use : $4R+3C$

Small circuit is needed to use : $3R+1C$

So $4R+3R \leq 100 \quad \therefore R \leq \dfrac{100}{7} \fallingdotseq 14.3$

and $2C+1C \leq 70 \quad \therefore C \leq \dfrac{70}{3} \fallingdotseq 23.3$, So the answer is 14.

QUESTION 45

Choice B

The sum of meeting and training time have to be no more than 16 hours per month.

So Meeting time(M) + Training time(T)$\leq$ 16 hours

There should be at least meeting time : $M \geq 2$

And There should be at least training time : $T \geq 1$

So when $M=2$, Maximum training time is when $T=14$

and when $M=2$, minimum training time is when $T=1$.

Therefore the difference between maximum training time and minimum training time is 13.

QUESTION 46

Choice A

Let first equation be Equ.① : $-5.1x+3y=1.2$

second equation be Equ.② : $3.2x-8y=b$

From Equ.①×8 + Equ.②×3 : $(-5.1\times 8+3.2\times 3)x = 12+b\times 3$

Since the coefficient of x is not zero, so these system of equation has one solution whenever b can be any number.

QUESTION 47

Choice C

Let first equation be Equ.① : $0.6=1.5\,(a+c\,(b+0.8))$

then $0.6 = 1.5a+1.5cb+1.2c$ ⋯ Equ.③

second equation be Equ.② : $-0.2=-2.5(b-0.4(1.2-1.5a))$

then $-0.2 = -2.5b+1.2-1.5a$⋯ Equ.④

From Equ.③ + Equ.④ : $0.4 = 1.5cb+1.2c-2.5b+1.2$

$(1.5c-2.5)b = -1.2c-1.2+0.4$

when the coefficient of b is zero, this equation can be no solution.

So $1.5c-2.5=0 \quad \therefore c=\dfrac{5}{3}$

QUESTION 48

Answer: 3

$$\sqrt{9} = \sqrt{3\times3} = \sqrt{3^2} = (3^2)^{\frac{1}{2}} = 3$$

QUESTION 49

Answer: 4

$$(\sqrt[3]{8})^2 = (\sqrt[3]{2^3})^2 = (2^{\frac{3}{3}})^2 = 2^2 = 4$$

QUESTION 50

Answer: $\frac{4}{5}$

$$\sqrt{\frac{16}{25}} = \sqrt{(\frac{4}{5})^2} = (\frac{4}{5})^{\frac{2}{2}} = \frac{4}{5}$$

QUESTION 51

Answer: 1

$$(\sqrt{3} - \sqrt{2})(\sqrt{3} + \sqrt{2}) = (\sqrt{3})^2 - (\sqrt{2})^2 = 3 - 2 = 1$$

QUESTION 52

Answer: 9

$\sqrt[3]{x-1} = 2$

$(x-1)^{\frac{1}{3}} = 2$

$x-1 = 2^3$

$\therefore x = 9$

QUESTION 53

Choice A

$$\frac{\sqrt{2}}{2}(\sqrt{8} - \sqrt{50}) = \frac{\sqrt{2}}{2}(2\sqrt{2} - 5\sqrt{2}) = (\sqrt{2})^2 - \frac{5\times(\sqrt{2})^2}{2} = 2-5 = -3$$

QUESTION 54

Choice A

$$\frac{8^{\frac{1}{2}}}{2^{\frac{1}{3}}}=\frac{(2^3)^{\frac{1}{2}}}{2^{\frac{1}{3}}}=2^{\frac{3}{2}-\frac{1}{3}}=2^{\frac{9}{6}-\frac{2}{6}}=2^{\frac{7}{6}}$$

QUESTION 55

Choice B

$$(2b^{-5})^3=\left(\frac{2}{b^5}\right)^3=\frac{2^3}{(b^5)^3}=\frac{8}{b^{15}}$$

QUESTION 56

Choice A

$$(16x^2)^{\frac{1}{2}}=16^{\frac{1}{2}}(x^2)^{\frac{1}{2}}=(4^2)^{\frac{1}{2}}|\,x\,|=4|\,x\,|$$

QUESTION 57

Choice B

$$\sqrt{x}\times\sqrt{\frac{y^5}{x^3}}=x^{\frac{1}{2}}\times\left(\frac{y^5}{x^3}\right)^{\frac{1}{2}}=x^{\frac{1}{2}}\times\frac{y^{\frac{5}{2}}}{x^{\frac{3}{2}}}=x^{\frac{1}{2}-\frac{3}{2}}\times y^{\frac{5}{2}}=x^{-1}\times y^{\frac{5}{2}}$$

QUESTION 58

Answer: $\frac{3}{4}$

$$\frac{\sqrt[5]{m^4}}{\sqrt[4]{m^3}}\div\frac{\sqrt{m}}{\sqrt[5]{m^6}}=\frac{m^{\frac{4}{5}}}{m^{\frac{3}{4}}}\times\frac{m^{\frac{6}{5}}}{m^{\frac{1}{2}}}=\frac{m^{\frac{10}{5}}}{m^{\frac{5}{4}}}=m^{2-\frac{5}{4}}=m^{\frac{3}{4}}$$

QUESTION 59

Choice B

$$\frac{\sqrt{x}}{5\sqrt[5]{x^6}} = \frac{x^{\frac{1}{2}}}{5 \times x^{\frac{6}{5}}} = \frac{1}{5}x^{\frac{1}{2}-\frac{6}{5}} = \frac{1}{5}x^{\frac{5}{10}-\frac{12}{10}} = \frac{1}{5}x^{-\frac{7}{10}} = \frac{1}{5x^{\frac{7}{10}}}$$

$$= \frac{1}{5\sqrt[10]{x^7}}$$

QUESTION 60

Choice A

Since $\sqrt{2x^2+7}+5 > 0$ for all real value of x, so the equation has no real solution.

QUESTION 61

Answer: 2

$$\frac{2x+6}{(x+2)^2} - \frac{2}{x+2} = \frac{2x+6}{(x+2)^2} - \frac{2(x+2)}{(x+2)^2} = \frac{2x+6-2x-4}{(x+2)^2} = \frac{2}{(x+2)^2}$$

$\therefore a = 2$

QUESTION 62

Choice A

Since $64b^2 - 25 = (8b)^2 - 5^2 = (8b+5)(8b-5)$

$2b + b^2 = b(2+b)$

$25 - 40b = 5(5-8b)$

so $$\frac{\frac{(64b^2-25)}{2b+b^2}}{\frac{25-40b}{b+2}} = \frac{\frac{(8b+5)(8b-5)}{b(2+b)}}{\frac{5(5-8b)}{b+2}} = \frac{(8b+5)(8b-5)(b+2)}{5b(b+2)(5-8b)} = -\frac{8b+5}{5b}$$

QUESTION 63

Choice B

Since $3x^2 - 2x - 5 = (3x-5)(x+1)$

$x^2 + x = x(x+1)$

$15x-9=3(5x-3)$

$$\therefore \frac{\dfrac{(3x^2-2x-5)}{x^2+x}}{\dfrac{15x-9}{3x}} = \frac{\dfrac{(3x-5)(x+1)}{x(x+1)}}{\dfrac{3(5x-3)}{3x}} = \frac{3x(3x-5)(x+1)}{3x(x+1)(5x-3)} = \frac{3x-5}{5x-3}$$

QUESTION 64

Answer: (1) $(3x-2)(x+4)$

(2) $(x^4+1)(x-2)$

(3) $(x^2-3)(x^3-2)$

(1) $3x^2-2x+12x-8=3x(x+4)-2(x+4)=(3x-2)(x+4)$

(2) $x^5+x-2x^4-2=x^4(x-2)+(x-2)=(x-2)(x^4+1)$

(3) $x^5-3x^3-2x^2+6=x^3(x^2-3)-2(x^2-3)=(x^3-2)(x^2-3)$

QUESTION 65

Answer: (1) $(x-3)(x+5)$

(2) $(x-4)(x-6)$

(3) $(x+3)^2$

(4) $(3x+2)(x-4)$

(5) $(5x-2)(x-3)$

(6) $2(2x-1)(x+3)$

(1) $x^2+2x-15=(x-3)(x+5)$

$x \quad 5$
$x \quad -3$

(2) $x^2-10x+24=(x-4)(x-6)$

$x \quad -4$
$x \quad -6$

(3) $x^2+6x+9=(x+3)^2$

$x \quad 3$
$x \quad 3$

(4) $3x^2+2x-8 = (3x+2)(x-4)$

$$\begin{matrix} 3x & \times & 2 \\ x & & -4 \end{matrix}$$

(5) $5x^2-17x+6 = (5x-2)(x-3)$

$$\begin{matrix} 5x & \times & -2 \\ x & & -3 \end{matrix}$$

(6) $4x^2+10x-6 = 2(2x^2+5x-3) = 2(2x-1)(x+3)$

$$\begin{matrix} 2x & \times & -1 \\ x & & 3 \end{matrix}$$

QUESTION 66

Choice B

$(v+\frac{1}{5})^2-9=0$

$(v+\frac{1}{5})^2=9$

$v+\frac{1}{5}=3$ or -3

$v=3-\frac{1}{5}$ or $-3-\frac{1}{5}$

$\therefore$ Sum of two roots $=-\frac{2}{5}$

QUESTION 67

Answer: (1) $-1\pm 2\sqrt{2}$

(2) -3

(3) $2\pm\sqrt{6}$

(4) 0 or 16

Use quadratic formula : If $ax^2+bx+c=0$, then $x=-\frac{b+\sqrt{b^2-4ac}}{2a}$

QUESTION 68

Choice A

$\text{LHS} = \frac{x}{x-1}+\frac{4}{x-2} = \frac{x(x-2)+4(x-1)}{(x-1)(x-2)} = \frac{x^2-2x+4x-4}{(x-1)(x-2)}$

$\text{RHS} = \frac{1}{(x-1)(x-2)}$

$$\therefore \frac{x^2-2x+4x-4}{(x-1)(x-2)} = \frac{4}{(x-1)(x-2)}$$

$\therefore\ x^2+2x-4 = 4$

$x^2+2x-8=0$

$(x+4)(x-2)=0$

$\therefore\ x = -4$ or 2

QUESTION 69

Choice B

From given the equation, $\dfrac{2x}{x+3} - \dfrac{4}{x-2} = -3$

If you multiply both sides by $(x+3)(x-2)$

$2x(x-2)-4(x+3) = -3(x+3)(x-2)$

$2x^2-4x-4x-12 = -3x^2-3x+18$

$5x^2-5x-30=0$

$x^2-x-6=0$

$(x-3)(x+2)=0$

$\therefore x=3$ or -2

$\therefore$ The sum of two roots $=1$

QUESTION 70

Choice A

$4+kp=4p^2$

$4p^2-kp-4= 0$

By the formulas of sum and products of roots on quardratic equation

Sum of roots $= -\dfrac{(-k)}{4} = 0$

$\therefore k=0$

QUESTION 71

Choice A

This problem is to find an equation in which the value of x does not exist when the value of y becomes 0.

QUESTION 72

Choice A

Price per ticket	Number of seats	The amount of money
25	4000	25×4000
26	3900	26×3900
27	3800	27×3800
…	…	…
$25+x$	$400-100x$	$(25+x)(400-100x)$

$\therefore y=(25+x)(400-100x)$

QUESTION 73

Choice D

The ball travels along the path of a parabola and reaches a maximum height of $10ft$ above ground level after traveling a horizontal distance of $5ft$. It means the vertex of coordinates is $(5, 10)$.

Since the vertex of standard form of quadratic function (h,k) is $y=a(x-h)^2+k$ where a, h and k are constants $(a\neq0)$, the quadratic form is $y=a(x-5)^2+10$.

And the initial position is $(0, 3)$.

$3=a(0-5)^2+10$

$\therefore a=-\dfrac{7}{25}$

$\therefore y=-\dfrac{7}{25}(x-5)^2+10$

QUESTION 74

Choice B

You have to find the form of quadratic that has the values of s when $d=0$.

The quadratic $-0.5(s-13)(s+14.2)$ can show the values of s when $d=0$.

QUESTION 75

Choice D

The volume of the cylinder is given by the equation

$$V(x)=20\pi(5+x)^2$$

where x is the additional length of the radius in centimeters.

You can calculate the present value of the volume of the cylinder when $x=0$.

$$V(0)=20\pi(5+0)^2 = 20\pi \times 25 = 500\pi$$

The quadratic $20\pi x^2+200\pi x+500\pi$ has the values of 500π.

QUESTION 76

Answer: $P = 100{,}000 \times 1.02^x$

Initial Population : $P_0 = 100{,}000$

The annual rate of interest : $r = 2\%$

The number of year : $n = x$

By the formula of the amount of money accumulated after n years, including interest, $P_0(1+\frac{r}{100})^n$

$\therefore P = 100{,}000 \times (1+\frac{2}{100})^x$

QUESTION 77

Choice C

From the formula, $d=15.69\,t^2$

when $\frac{4}{9}d$ is replaced instead of d, $\frac{2}{3}t$ have to be replaced instead of t in order to establish an equality. So, $\frac{2}{3}$ of the time it required to pass the cone.

QUESTION 78

Choice D

Initial Population of lemna minor : $P_0 = 10$

The annual rate of interest : $r = 100\%$ (double)

The number of increasing period for d days : $n = \frac{d}{2}$ (double every two days)

By the formula of the amount of money accumulated after n years, including interest, $P_0(1+\frac{r}{100})^n$, the area of lemna minor $f(d)$ is

$\therefore f(d)=10\times(1+\frac{100}{100})^{\frac{d}{2}}$

QUESTION 79

Choice A

Black tea is prepared by pouring boiling water (100 $°C$ onto tea leaves and allowing the tea to brew in a pot or cup.

Since in a room whose temperature is 20 $°C$ and the tea reaches a temperature of 60 $°C$ after about 4 minutes, this function model is a decreasing exponential function with a horizontal asymptote, $T = 20\ °C$.

QUESTION 80

Choice D

The temperature T in degrees Celsius of a chilled drink after m minutes sitting on a table is given by the following function. $T(m) = 32 - 28 \cdot 3^{-0.05m}$

Since after a very long time, m will be a big number. It means the drink will warm up to 32 degrees Celsius.

QUESTION 81

Choice A

Since the function $p(t)$ is a polynomial of t such that $(t-10)$, $(22-t)$, $(t+10)$, and $(20+t)$ are all factors of $p(t)$, The graph of $y = p(t)$ in the ty-plane has four t intercepts or more including $t = 10$, $t = 22$, $t = -10$ and $t = -20$.

QUESTION 82

Choice D

The fact that a polynomial function has a double zero means the graph of polynomial function has a touching point on x axis.

QUESTION 83

Answer: 54

Since Given $w(x) = ax^2 + bx + c$, $a = 3$ and $w(3) = w(15) = 0$,

$w(x) = 3x^2 + bx + c = 3(x-3)(x-15)$

$\therefore 3x^2 + bx + c = 3x^2 - 54x + 135$

$\therefore b = -54$

$\therefore |b| = 54$

QUESTION 84

Answer: 0

$h(x) = (x-a) \times s(x)$

$= (x-a)(x-4)(x-5)^2$

$= (x-a)^3 Q(x) \qquad \therefore a = 5$

$\therefore \ s(a) = s(5) = (5-4)(5-5)^2 = 0$

QUESTION 85

Choice C

$y = w(x)$

$\downarrow$ translation of $\begin{pmatrix} -3 \\ +4 \end{pmatrix}$

$y = p(x)$

$\downarrow$ translation of $\begin{pmatrix} +3 \\ -4 \end{pmatrix}$

$y + 4 = p(x-3)$

$\therefore y = p(x-3) - 4$

$\therefore w(x) = p(x-3) - 4$

QUESTION 86

Answer: 2

$y = -f(-x)$

$\Leftrightarrow f(-x) = -y$

You have to find the value of $f(x) = 1$ and from given graph, $f(-2) = 1$.

$\therefore \ -x = -2$ and $-y = -1$

$\therefore x = 2$

QUESTION 87

Choice B

$y = g(x)$

$\downarrow$ Vertical stretching factor : 3

$y = 3g(x)$

$\downarrow$ Reflected over y axis

$y = 3g(-x)$

QUESTION 88

Answer: $\dfrac{59}{9}$

$9x-10y=19$

↓ translation of 4 units downword

$9x-10(y+4)=19$

when $y=0$,

$9x-10(0+4)=19$

$9x-40=19$

$\therefore x=\dfrac{59}{9}$

QUESTION 89

Choice B

$y=\dfrac{1}{x-3}+0$ have a vertical asymptote with the equation $x=3$ and a horizontal asymptote with the equation $y=0$.

QUESTION 90

Choice B

$y=\dfrac{3}{x+4}+0$ have a vertical asymptote with the equation $x=-4$ and a horizontal asymptote with the equation $y=0$.

And $y=\dfrac{3}{x+4}-3$ have a vertical asymptote with the equation $x=-4$ and a horizontal asymptote with the equation $y=-3$.

So, both graphs have a vertical asymptote at $x=-4$.

QUESTION 91

Choice C

Since the rational function f is defined by an equation in the form $f(x)=\dfrac{a}{x+b}$, where a and b are constants and $y=f(x)$ has passed two points $(-10,-1)$ and $(-7,-2)$

$y=f(x)=\dfrac{a}{x+b}$

$(-10, -1)$: $-1 = \dfrac{a}{-10+b}$ $\quad \therefore a = 10 - b$

$(-7, -2)$: $-2 = \dfrac{a}{-7+b}$ $\quad \therefore a = 14 - 2b$

$\therefore 10 - b = 14 - 2b$

$\therefore b = 4,\ a = 6$

$\therefore f(x) = \dfrac{6}{x+4}$

$\therefore y = g(x) = f(x+4) = \dfrac{6}{(x+4)+4}$

$\therefore g(x) = \dfrac{6}{x+8}$

QUESTION 92

Choice B

The function $f(x) = \sqrt{x-7}$ has the value of $x=7$ when $y=0$. So the graph of the function intersects the x-axis at (7,0).

QUESTION 93

Answer:

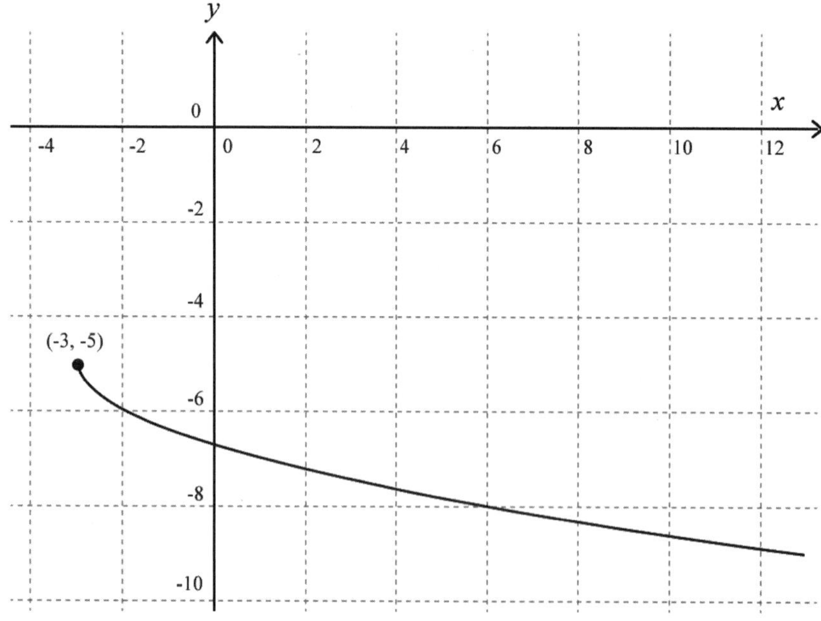

$y = \sqrt{x}$

$\downarrow$ Reflected over x axis

$y = -\sqrt{x}$

$\downarrow$ translation of $\begin{pmatrix} -3 \\ -5 \end{pmatrix}$

$y+5 = -\sqrt{x+3}$

QUESTION 94

Choice C

As $c = \dfrac{ab}{a+b}$

$c(a+b) = ab$

$ca + cb = ab$

$a(b-c) = cb$

$\therefore a = \dfrac{cb}{b-c}$

QUESTION 95

Choice C

Given that $x = v_0 t + \dfrac{1}{2}at^2$

$x - v_0 t = \dfrac{1}{2}at^2$

$2(x - v_0 t) = at^2$

$\therefore a = \dfrac{2(x - v_0 t)}{t^2}$

QUESTION 96

Choice C

742 words : 14 minutes = x words : 60 minutes

$14x = 742 \times 60$

$\therefore x = 3180$

QUESTION 97

Choice A

350 males = 150 (with eggs) + 200 (without eggs)

150 (with eggs) : 200 (without eggs) = 3 : 4

QUESTION 98

Choice D

1 gallon : \$4 : 25 miles

$\downarrow \div 4$

$\frac{1}{4}$ gallon : \$1 : $\frac{25}{4}$ miles

$\downarrow \times 5$

$\frac{5}{4}$ gallon : \$5 : $\frac{25}{4} \times 5$ miles

$\therefore \frac{25}{4} \times 5$ miles $= m$ miles

$\therefore \frac{4}{25} \times m$ miles $= 5$ miles

QUESTION 99

Choice A

36 pounds (lb)

$\downarrow$ (35% $\downarrow$) or $\times 0.65$ after trim

36×0.65

$\downarrow$ (75% $\downarrow$) or $\times 0.25$ after cooking

$36 \times 0.65 \times 0.25 = 5.85$

QUESTION 100

Choice B

8 minutes : 35% $= x$ minutes : 65%

$35x = 8 \times 65x = \dfrac{8 \times 65}{35} = 14.85 \fallingdotseq 15$ minutes

QUESTION 101

Choice B

$$\frac{27 \times 125}{1{,}000{,}000} \times 100\,(\%) = 0.0152\,(\%)$$

QUESTION 102

Choice B

Last year : $\frac{406}{452} \times 100(\%) = 89.82(\%)$

$\downarrow +2\%$

This year : $\frac{x}{436} \times 100(\%) = 89.82(\%) + 2(\%)$

$\therefore \ x = \frac{91.82(\%) \times 436}{100} = 400.348 \fallingdotseq 400$

QUESTION 103

Answer : 1.5

Let's focus on the purely(100%) amount of saline per each solutions and mixed solutions.

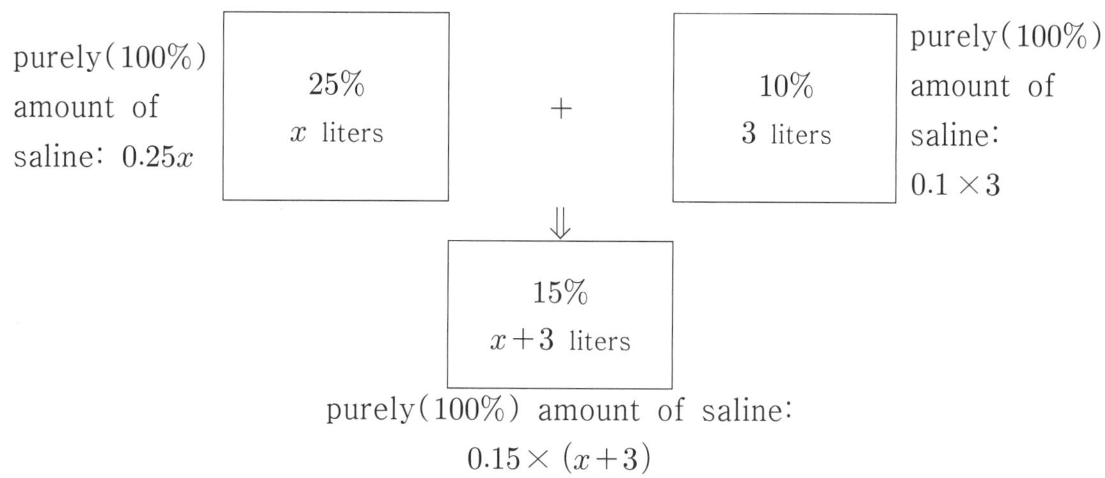

$\therefore 0.25x + 0.3 = 0.15(x+3)$

$0.1x = 0.45 - 0.3$

$\therefore x = 1.5$

QUESTION 104

Answer : 2

With respects to volumes

Small Box (2 paper books) : Larger Box (16 paper books)

$x^3 : (kx)^3 = 2 : 16$

$2k^3x^3 = 16x^3$

$k^3 = 8$

$\therefore k = 2$

QUESTION 105

Answer : 588

Density = mass / volume = 807 (Density of a sampke of oak) = $\dfrac{x\,kg}{0.9 \times 0.9 \times 0.9\,m^3}$

$\therefore x = 807 \times 0.9^3 = 588.303 \fallingdotseq 588$

QUESTION 106

Choice B

Since 1 mile =1760 yards

1^2 mile^2 =1760^2 yards^2

$\therefore$ 1^2 mile^2 : 1760^2 yards^2 = x mile^2 : 11,863,808 yards^2

$$x = \frac{11{,}863{,}808}{1760^2}$$

$\therefore$ $x = 3.83$ mile^2

QUESTION 107

Choice D

Since 1 mile =1760 yards

1^2 mile^2 =1760^2 yards^2

$\therefore$ 4.36 mile^2 = 4.36 × 1760^2 yards^2

= 13,505,536 yards^2

QUESTION 108

Choice A

As the problem is given, a watt is equivalent to 1 joule per second $\left(\dfrac{j}{s}\right)$

6.5×10^2 kilojoules per centisecond $\left(\dfrac{kj}{cs}\right) = 6.5\times 10^2 \times \dfrac{10^3}{10^{-2}}$ joule per second $\left(\dfrac{j}{s}\right)$

$= 6.5\times 10^{2+3-(-2)} \left(\dfrac{j}{s}\right)$

$= 6.5\times 10^7 \left(\dfrac{j}{s}\right) = 6.5\times 10^7$ watts

QUESTION 109

Choice D

1 lux = 1 candelas / m^2

$\therefore$ 0.0002 kilo candelas / cm^2

$= 0.0002 \times 10^3$ candelas / 10^{-4} m^2

$= 0.0002 \times 10^7$ candelas / m^2

= 2000 lux

QUESTION 110

Choice C

Note 1 mile = $1.6km = 1600$ meters

Speed of cheetah $= \dfrac{100m}{5.96\,\text{sec}}$

$\therefore$ $100m : 5.95\,\text{sec}s = x\,m : 1$ hour

$100m : 5.95\,\text{sec}s = x\,m : 3600$ seconds

$\therefore$ $x = \dfrac{3600 \times 100}{5.95}$ m/hour (moving distance per hour)

Since 1 mile = $1.6km = 1600$ meters ,

1 mile : 1600 meters $= y$ mile : $\dfrac{3600 \times 100}{5.65}$ meters

$\therefore$ $1600 \times y = \dfrac{3600 \times 100}{5.65}$

$\therefore y \fallingdotseq 37.82$

QUESTION 111

Choice D

The probability that Victor took late (6) when he used Route A (11)

$= \dfrac{6}{11}$

QUESTION 112

Choice B

Spanish non−native speakers : all non−native speakers $= \dfrac{x}{1273} = 4.8(\%) = 0.048$

$\therefore\ x = 64.104 \fallingdotseq 61$

QUESTION 113

Choice D

Let the number of people that 'not want sauce and want cheese'

	Cheese	No cheese	Total
Sauce	−	−	18
No sauce	x	−	8
Total	20	6	26

P(Not want sauce | given that want cheese) $= \dfrac{1}{5}$

$\dfrac{x}{20} = \dfrac{1}{5}$

$\therefore x = 4$

Then you can fill all numbers on the table below.

	Cheese	No cheese	Total
Sauce	16	2	18
No sauce	$x=4$	4	8
Total	20	6	26

$\therefore P$(want cheese | given that want sauce) $= \dfrac{16}{18} = \dfrac{8}{9}$

QUESTION 114

Answer : 12

By the given conditions, you can fill all numbers on the table below.

	Fewer than 200 pages	200 pages and above	Total
19th Centry	two books are from 19th century and fewer than 200 pages $=2$	−	−
20th Centry	−	−	14
Total	4	20	−

So

	Fewer than 200 pages	200 pages and above	Total
19th Centry	two books are from 19th century and fewer than 200 pages = 2	8	10
20th Centry	2	12	14
Total	4	20	24

∴ The number of books, both from 20th century that are 200 pages and above = 12

QUESTION 115

Choice B

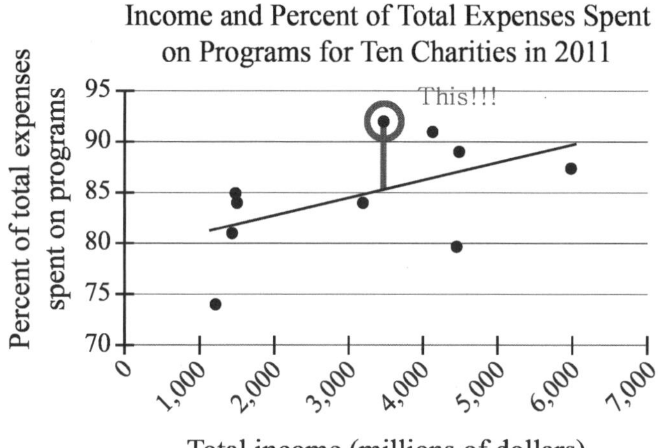

On the given diagram you can estimate the difference of the actual percent (Notated 'THIS!!') and its predict percent on the line.

QUESTION 116

Choice B

On the given diagram you can estimate the slope of the sales of organic food is about $\frac{5}{2}$. $y = 15.1 + 2.4x$ has the most closest value of this slope.

QUESTION 117

Choice C

On the given data table you can estimate the slope of $280-293$(in thousands) every 5 years, $\dfrac{280-293}{5}=\dfrac{-13}{5}$. And the number of years since 1986 ($t=0$) is 293 (in thousands). So the most similar function is $f(t)=-2.94t+293$.

QUESTION 118

Answer : 3

When $t=3$

$P=215\,(1.005)^{\frac{t}{3}}=215\,(1.005)^{\frac{3}{3}}=215\,(1.005)^{1}$

This means the population is predicted to increase by 0.5% every 3 months.

QUESTION 119

Choice D

According to the given scatterplot, China's production of greenhouse products are growing exponentially, but and Canada's is growing linearly.

QUESTION 120

Choice A

A function that models the data shown is $y=4.19(1.07)^{x}$ and the value 4.19 in the model is the value when $x=0$. It means it will take approximately 4.19 grams of water vapor to saturate the air when the air temperature is 0 degrees Celsius.

QUESTION 121

Choice D

Since $c(x)=110\times1.103^{x}=110\times(1+\dfrac{10.3}{100})^{x}$

It means, between 1970 and 2010, the cost to run a 30 seconds advertisement during this sporting event increased by about 10.3%(about 10%) each year.

QUESTION 122

Choice C

The given scatterplot is a increasing exponential model. So you can have to select Choice C or D because their base number is more than 1. And when $t=70$, $P \fallingdotseq 7$.

On Choice C : $0.53 \times 1.035^{70} \fallingdotseq 5.8897$

But On Choice D : $0.53 \times 1.1410^{70} = 1.47 \times 10^{10}$ (Too bigger than $P=7$)

QUESTION 123

Choice B

The given scatterplot is a decreasing exponential model. So you can have to select Choice B or D because their base number is less than 1.

Since when $x=0$, $f(x) \fallingdotseq 3000$, the exponential model with the most similar initial values is B.

QUESTION 124

Choice C

Average rate of change before 35

= Slope between two points (10, 1591) and (35, 3873)

$= \dfrac{3873-1591}{35-10} = \dfrac{2282}{25} \fallingdotseq 91.28$

This means The company charges about $91 per terabyte after the tenth terabyte but before the thirty-fifth.

QUESTION 125

Choice B

According to the graph,

The total cost of clothing factory producing 20 shirts is $70.

It means the cost of clothing factory producing 1 shirts is $\$\dfrac{70}{20} = \$3.5$.

But the total cost of clothing factory producing 80 shirts is $100.

It means the cost of clothing factory producing 1 shirts is $\$\dfrac{100}{80} = \$1.25$.

Therefore we can tell the average cost always decreases as the number of shirts produced in one hour increases.

QUESTION 126

Choice A

A tankless water heater and tank water heater are \$4,090 each at the end of the 7th year of use. ⇒ Initial values : \$4,090

The cumulative costs of the tankless heater increase \$180 per year for 3 years

$\Rightarrow\ 4090 + 180 \times 3$

but for the tank heater, they increase by 11% per year.

$\Rightarrow\ 4090 \times (1 + \frac{11}{100})^3$

$\therefore\ 4090 \times (1 + \frac{11}{100})^3 - (4090 + 180 \times 3) = 963.61079 \fallingdotseq 964$

QUESTION 127

Answer : 384

There were 6000 genetics articles indexed at the end of that time. ⇒ Initial values : 6,000

In a particular science research database, the number of indexed genetics articles had increased by an average of 307 articles per year for two years.

$\Rightarrow\ 6000 + 307 \times 2$

The number of articles increased by 8% annually since then.

$\Rightarrow\ 6000 \times (1 + \frac{8}{100})^2$

$\therefore\ 6000 \times (1 + \frac{8}{100})^2 - (6000 + 307 \times 2) = 384.2 \fallingdotseq 384$

QUESTION 128

Choice B

A state accountant models that sales tax revenue grew about 15% per year for several years. Then 2 years ago, when the revenue was \$2.1 billion, the revenue began growing by \$179.5 million per year instead. About how much less would the revenue be this year than if it had continue d growing 15% per year?

$\Rightarrow\ (2.1 \times 1000) \times (1 + \frac{15}{100})^2 - (2.1 \times 1000 + 179.5 \times 2) = \$318.25$ millions

QUESTION 129

Answer : 18

According to the given conditions of this problem

$$1000\times(1+\frac{x}{100})^3-(800+200\times3)=240$$

$$(1+\frac{x}{100})^3=1.64$$

$$1+\frac{x}{100}=1.64^{\frac{1}{3}}$$

$$\therefore x=17.92\fallingdotseq 18$$

QUESTION 130

Choice D

$$f(t)=\frac{f(t-1)}{1.004}=f(t-1)\times0.9956=f(t-1)\times(1-0.00438)$$

It means the rate of water flow decreases exponentially.

QUESTION 131

Choice A

According to the data table, it is approximately linear because the concentration increases by about the same amount, 60 $(\frac{\mu mol}{L})$ for each $50nA$ increase in current.

QUESTION 132

Answer : 289

According to the given conditions of this problem,

$$500\times1.2^3-(500+25\times3)=289$$

QUESTION 133

Choice A

The average rate of increase in temperature per second between $s=0$ and $s=15$

$$=\frac{1650-0}{15-0}=110$$

QUESTION 134

Choice D

The given scatterplot is a concave downward quadratic model. So you can have to select Choice B or D. And since $x=0$, $y \fallingdotseq 750$, the quadratic model with the most similar initial values is D.

QUESTION 135

Choice B

If the outlier is removed, sum of all data is very smaller. so the value of the mean of the data set will decrease.

QUESTION 136

Answer : 6.1

$$\text{Mean} = \frac{5\times 9+7\times 11}{20} \fallingdotseq 6.1$$

QUESTION 137

Answer : 20

$$\text{Mean} = \frac{3\times 25+2\times 20+2\times 20+3\times 15}{3+2+2+3} = 20$$

QUESTION 138

Choice B

Tank A $<$ 5 ounces	5 ounces $<$ Tank B $<$ 13 ounces	Tank C $>$ 13 ounces
10 fish	11 fish	13 fish
Total 33 fish		

$\therefore 5 \le$ median ≤ 13

QUESTION 139

Choice A

As a result, the average number of dwelling residents decreases because the absolute number of dwelling residents decreases.

QUESTION 140

Choice B

There are 30 customers. So median is the mid term between 15th and 16th data.

$$\therefore \text{ Median } = \frac{4+5}{2} = 4.5$$

QUESTION 141

Choice D

Since the data in beats per minutes after exercise is more spreading, $s_1 > s_2$

And

Range of beats per minutes before exercise : $88 - 56 = 32$

Range of beats per minutes after exercise : $112 - 80 = 32$

$\therefore r_1 = r_2$

QUESTION 142

Choice A

The range of first data set, the worm gummy distances, has more small and the data is concentrated. so the worm gummy distances have a greater standard deviation and mean than the fish gummy distances.

QUESTION 143

Choice C

There is a gap in the middle of the second graph. This means that there are more data that differ from the overall average value.

The standard deviation for the 2nd shift employees is greater.

QUESTION 144

Choice B

The data in the first graph, class A are more centralized. In the second graph, class B, the frequency of the central data and the surrounding data is similar. It means The standard deviation of the scores from class B is greater than the standard deviation of the scores from class A.

QUESTION 145

Answer : 4, 5 or 6

In year 4, the difference between the actual revenue and the predicted revenue is about 5.

QUESTION 146

Choice C

In general, the larger the size of the sample, the larger the margin of error.

QUESTION 147

Choice D

$0.94-0.12 \leq$ Estimated actual values of parameter $\leq 0.94+0.12$

QUESTION 148

Choice D

It can be expected that there is a correlation between the two data, but it cannot be said that forest decline definitely reduces brown bear population.

QUESTION 149

Choice B

This is all you can guess from the given graph, By the age of 20, the best experts and professional violinists in the study had practiced more than twice as much as the least accomplished violinists.

QUESTION 150

Choice D

Samples with specific conditions that are not common are not good samples of an appropriate survey process like enjoying the weather portion or members of the local meteorological society.

QUESTION 151

Choice B

This is a question of asking examples of systematic samples. So is the systematic sample. Samples with specific conditions that are not common are not good samples of an appropriate survey process like there should be no conditions related to the library, a football team, or a specific grade.

QUESTION 152

Choice D

Because the absolute amount of figures cannot be determined by the ratio alone, we can only know the following facts : Viewers voting by social media were likely to prefer Contestant 2 than more viewer voting by text message.

QUESTION 153

Choice D

Case 1 : $\angle A = 50, \angle B = 65, \angle C = 65$

Case 2 : $\angle A = 50, \angle B = 50, \angle C = 80$

QUESTION 154

Choice D

$$\frac{7\pi k^3}{48} = 473$$

$$k^3 = \frac{473 \times 48}{7\pi}$$

$$\therefore k = \left(\frac{473 \times 48}{7\pi}\right)^{\frac{1}{3}} \fallingdotseq 10.11$$

QUESTION 155

Choice D.

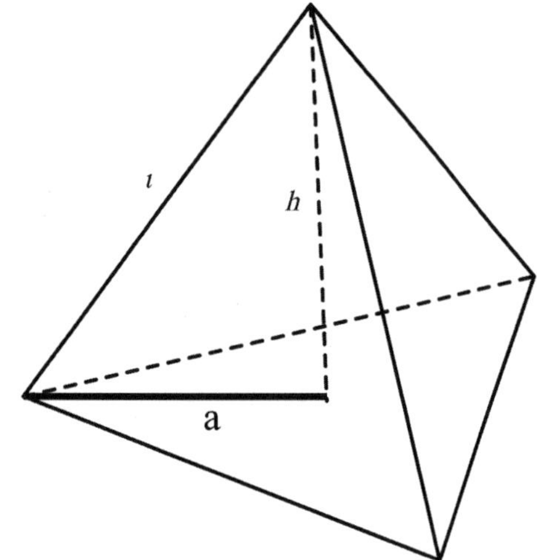

From the diagram above,

$$a = \frac{l}{2} \times \sqrt{3} \times \frac{2}{3} = \frac{\sqrt{3}}{3}l$$

And By Pythagorean theorem, $a^2 + h^2 = l^2$

$$\therefore h^2 = l^2 - a^2$$

$$= l^2 - \left(\frac{\sqrt{3}}{3}l\right)^2 = \frac{2}{3}l^2$$

$$\therefore h = \sqrt{\frac{2}{3}}l = \frac{\sqrt{6}}{3}l$$

QUESTION 156

Choice A

$$V = \frac{1}{3} \times \frac{\sqrt{3}}{4}l^2 \times \frac{\sqrt{6}}{3}l = \frac{\sqrt{16}}{36}l^3 = \frac{3\sqrt{2}}{36}l^3 = \frac{\sqrt{2}}{12}l$$

QUESTION 157

Choice A

Volume = Sphere + cylinder

$$= \frac{4}{3}\pi \times 3^3 + \pi \times 3^2 \times 10$$

$$= 36\pi + 90\pi = 126\pi$$

QUESTION 158

Choice B

Volume = Big cylinder - internal cylinder

$= \pi \times 3^2 \times 30 - \pi \times 2^2 \times 30 = 150\pi$

QUESTION 159

Choice A

Let the radius of the Buckyball, $2\pi r = 70 \ \therefore r = \frac{35}{\pi}$

So volume of the Buckyball $= \frac{4}{3}\pi \times (\frac{35}{\pi})^3 \fallingdotseq 5{,}792$

QUESTION 160

Choice D

Volume of 'conical bottom tank',(V)= cylinder + Cone

$V = \pi \times 60^2 \times 125 + \frac{1}{3} \times \pi \times 60^2 \times 50 \fallingdotseq 2{,}167{,}700$

QUESTION 161

Choice B

Let the length of interior right square pyramid, x

$x^2 + x^2 = 40^2$

$\therefore x = \frac{40}{\sqrt{2}}$

$\therefore$ Volume $= \frac{1}{3} \times (\frac{40}{\sqrt{2}})^2 \times 20 \fallingdotseq 5333$

QUESTION 162

Choice B

By special right angled triangle(30°, 60°, 90°) ratios,

If $PQ = 1$, then $QT = \frac{1}{2}$, $TR = \frac{1}{4}$ and $RT = \frac{1}{4} \times \frac{1}{\sqrt{3}} = \frac{\sqrt{3}}{12}$

QUESTION 163

Choice B

$AB = \sqrt{3^2+4^2} = 5$

And $BC = 2AB = 10$ (Given)

For example, in the case of choice A, The distance between $(3, 4)$ and $(-5, -2)$:

$\sqrt{(-5-3)^2+(-2-4)^2} = \sqrt{64+36} = 10$.

But in the case of choice B The distance between $(3, 4)$ and $(9, 12)$:

$\sqrt{(9-3)^2+(13-4)^2} = \sqrt{36+81} \neq 10$

Since the length of line segment BC is twice the length of segment AB, the point could not be the coordinates of point C, so the answer is Choice B.

QUESTION 164

Answer : 15

Let the base of the stage be $2b$, then since depth 8 meters (m) and two walls each of length 10 meters (m), $b^2+8^2=10^2$. $\therefore b=6$ By Pythagorean theorem.

As the seating portion of the hall has an area of 180 square meters,

$(2b)\times x = 12\times x = 180 \quad \therefore x = 15$

QUESTION 165

Answer : 42

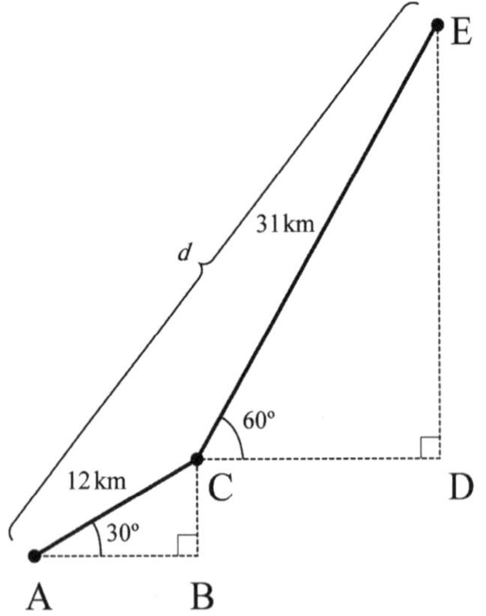

By special right angled triangle($30°, 60°, 90°$) ratios,

$AB = 6\sqrt{3}$, $BC = 6$, $CD = \dfrac{31}{2}$ and $DE = \dfrac{31\sqrt{3}}{2}$

By Pythagorean theorem

$\therefore d^2 = (AB + CD)^2 + (BC + DE)^2$

$\therefore d = \sqrt{(6\sqrt{3} + \frac{31}{2})^2 + \left(6 + \frac{31\sqrt{3}}{2}\right)^2} \fallingdotseq 42$

QUESTION 166

Choice C

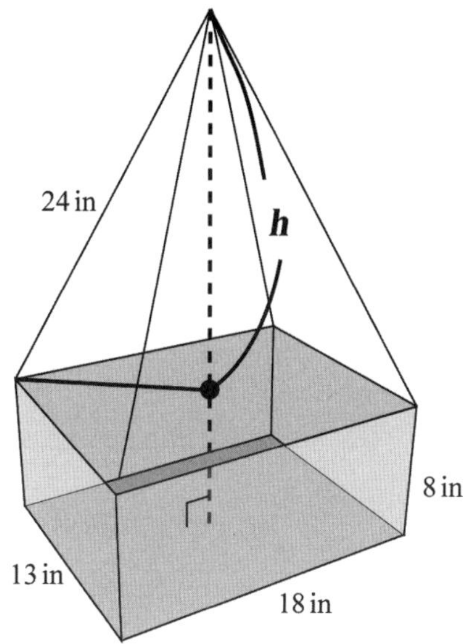

The length of diagonal of rectangular base $= \sqrt{13^2 + 18^2}$

$\therefore h^2 = \sqrt{24^2 - (\frac{\sqrt{13^2 + 18^2}}{2})^2}$

$\therefore$ Total height of the hanging planter $= h + 8 \fallingdotseq 29.28$

QUESTION 167

Answer : 16

By special right angled triangle(30°, 60°, 90°) ratios,

$CA = 16$, $CD = 16\sqrt{2} = x\sqrt{2}$

$\therefore x = 16$

QUESTION 168

Choice B

A rectangle placed in $\triangle ABC$ by the median theorem of the triangle is a parallelogram. A two inside opposite angles of parallelogram has the same size.

$\therefore \beta = t$

QUESTION 169

Choice A

Since the sum of alternating angles between lines w and x is $180°(=96°+84°)$.

$\therefore$ Two lines w and x is parallel.

QUESTION 170

Answer : 6

Since $\tan B = \dfrac{3}{4}$ in the figure, let $AB = 4x$ and $AC = 3x$.

On Right $\triangle ABC$, $(4x)^2 + (3x)^2 = 15^2$ By Pythagorean theorem

$\therefore x = 3$

And since $DA = 4$, $DB = 4x - 4 = 4 \times 3 - 4 = 8$.

On Right $\triangle BDE$, $\tan B = \dfrac{3}{4} = \dfrac{DE}{DB}$.

$\dfrac{3}{4} = \dfrac{DE}{8}$ $\quad \therefore DE = 6$

QUESTION 171

Answer : 31

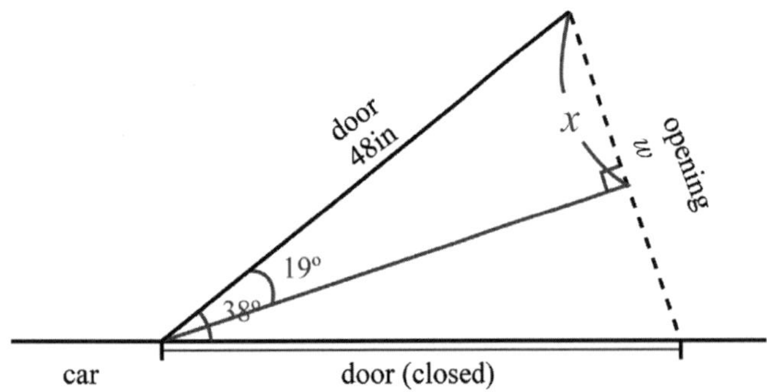

In the figure above, $\dfrac{x}{48} = \sin 19°$

$\therefore x = 48 \times \sin 19°$

$\therefore w = 2x = 31.252 \cdots \fallingdotseq 31$

QUESTION 172

Choice B

$\cos(90-x)^\circ = \sin x^\circ = 0.9$

QUESTION 173

Answer : 42

Since $\angle ZYX = \angle XYZ = \theta$

$\cos\theta = \dfrac{36}{x} = \dfrac{x}{49}$

$\therefore x^2 = 36 \times 49$

$\therefore x = 42$

QUESTION 174

Answer : 40

Let $CD = x$, $BD = y$ then $AD = \sqrt{3}y$

$\dfrac{y}{20} = \sin 65^\circ \quad \therefore y = 20\sin 65^\circ$

$\dfrac{x}{20} = \cos 65^\circ \quad \therefore x = 20\cos 65^\circ$

$\therefore AC = \sqrt{3}y + x = \sqrt{3} \times 20\sin 65^\circ + 20\cos 65^\circ = 39.847\cdots$

$\therefore AC \fallingdotseq 40$

QUESTION 175

Choice B

Since π radian $= 180^\circ$

$\therefore \ \theta = \dfrac{\pi}{2}$radian $= 90^\circ$

QUESTION 176

Choice A

Arc length $s = 2 \times \pi \times 6 \times \dfrac{220}{360} = 23.038 \fallingdotseq 23$

QUESTION 177

Choice C

Arc length $=20=2\times\pi\times4\times\dfrac{x}{360}$

$\therefore x=\dfrac{20\times360}{8\pi}=286.478\cdots$

$\therefore x\fallingdotseq286$

QUESTION 178

Choice B

Since $\triangle ABC$ is a right angled triangle and $\angle A=30^\circ$, $\angle B=90^\circ$ and $\angle C=60^\circ$. By special right angled triangle($30^\circ, 60^\circ, 90^\circ$) ratios, If $AB=6$,

then $BC=\dfrac{6}{\sqrt{3}}$.

$\therefore$ Area of $\triangle ABC=\dfrac{1}{2}\times6\times\dfrac{6}{\sqrt{3}}=\dfrac{36}{2\sqrt{3}}=\dfrac{18}{\sqrt{3}}=6\sqrt{3}$

QUESTION 179

Choice C

Let the length of radius of Earth be r.

Then $2\pi r=40{,}030$

$\therefore 2r=\dfrac{40{,}030}{\pi}$

$1\,km$: 0.62137 miles $=\dfrac{40{,}030}{\pi}km$: x miles

$\therefore x=0.62137\times\dfrac{40{,}030}{\pi}=7917.46\cdots\fallingdotseq7{,}917$

QUESTION 180

Answer : 2.1

Since the radius of each of her old tires is 0.30 meter, and the radius of her old tires is 0.30 meter, and the radius of each of her new tires is 11% larger than the radius of one of her old tires $2\pi\times0.3+2\pi\times0.3\times0.11=2.0923\fallingdotseq2.1$.

QUESTION 181

Answer : 12.5

Let the length of radius $= r$.

then $WO = r$, and $ZO = 9 - r$.

On $\triangle OXZ$

$12^2 + (9-r)^2 = r^2$

$12^2 + 81 - 18r + r^2 = r^2$

$18r = 144 + 81$

$\therefore r = 12.5$

QUESTION 182

Answer : 1.2

Since the diameter of the circle is 30, the length of radius is 5.

when the arc YXZ has length 18 and the measure, a radian, of angle YOZ,

By the formulas $l = r\theta$

$18 = 15 \times a$

$\therefore a = 1.2$

QUESTION 183

Answer : 17.5

Area of sector $= \frac{1}{2} \times 5 \times 7 = \frac{35}{2} = 17.5$

QUESTION 184

Choice D

Area of sector $= 500 = \frac{1}{2} \times$length of radius$\times$arc length $= \frac{1}{2} \times l \times 40$

$\therefore l = 25$

QUESTION 185

Choice B

On rectangle $ABCR$, $AC = BR = 6$.

And since $AC + RC = 8$ $\quad \therefore RC = 2$

On $\triangle ACR$, $AR^2 + 2^2 = 6^2$ $\quad \therefore AR = \sqrt{32}$ $\;\therefore AS = RS - AR = 6 - \sqrt{32}$

$\therefore$ The perimeter of shaded region $=$ Arc $SBT + TC + AC + AS$

$$= \frac{1}{4} \times 2 \times \pi \times 6 + 4 + 6 + (6 - \sqrt{32})$$

$$= 3\pi + 10.343 \fallingdotseq 3\pi + 10$$

QUESTION 186

Choice A

The radius of circle $(x+3)^2 + (y-1)^2 = 25$ is 5, and its center coordinate is $(-3, 1)$. The length between the point that lies in the interior of the circle and the center coordinate is less than 5.

QUESTION 187

Answer : 8

$2x^2 + 2y^2 - 8x - 6y - 16 = 0$

$x^2 + y^2 - 4x - 3y - 8 = 0$

$(x-2)^2 + (y - \frac{3}{2})^2 = 4 + \frac{9}{4} + 8$

$\therefore r = \sqrt{4 + \frac{9}{4} + 8} = 3.774\cdots$

$\therefore 2r = 7.549 \fallingdotseq 8$

QUESTION 188

Answer : 8

$x^2 + y^2 - 6x - 10y = 2$

$(x-3)^2 + (y-5)^2 = 2 + 9 + 25$

The center coordinate : $(3, 5) = (a, b)$

$\therefore\ a + b = 8$

QUESTION 189

Choice D

If a circle in the xy-plane has a center at $(\frac{5}{8}, -\frac{5}{6})$ and a length of radius of $\frac{7}{20}$, the equation of the circle is

$\therefore \left(x-\frac{5}{8}\right)^2+\left(y+\frac{6}{5}\right)=\frac{49}{400}$

QUESTION 190

Choice D

The center coordinate that The diameter of a circle graphed in the xy-plane has endpoints at $(-23, 15)$ and $(1, -55)$ is a midpoint between two points.

$\therefore$ The midpoint between two points $(-23, 15)$ and $(1, -55)$

$$=\left(\frac{-23+1}{2}, \frac{15-55}{2}\right)=(-11, -20)$$

$\therefore$ The length of radius = The distance between $(-23, 15)$ and $(-11, 20)$

$\therefore r=\sqrt{(-23+11)^2+(15-(-20))^2}$

$\therefore r^2=1369$

$\therefore (x+11)^2+(y+20)^2=1369$

QUESTION 191

Answer : 25 or -15

$x^2+y^2-10x+34y-527=0$

$(x-5)^2+(y+17)^2=527+5^2+17^2=841$

$r=\sqrt{841}=29$

Since the distance between $(5, 17)$ and $(x, -38)$ is 29,

$29=\sqrt{(x-5)^2+(-38+17)^2}$

$29^2=(x-5)^2+21^2$

$(x-5)^2=29^2-21^2=400$

$\therefore x-5=20$ or -20

$\therefore x=25$ or -15

QUESTION 192

Choice D

$\sin\left(\frac{\pi}{2}\right)=\sin(90^\circ)=1$

QUESTION 193

Answer : $\frac{4}{5}$

$\cos(90-x)^\circ = \sin x^\circ = \frac{4}{5}$

QUESTION 194

Choice B

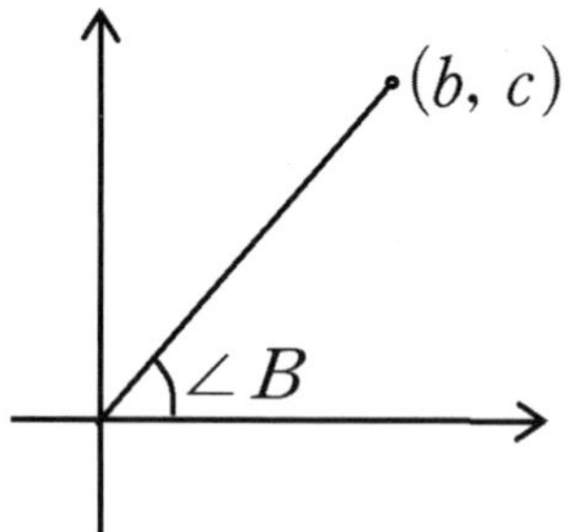

Given a point (b, c) on xy plane above, by definition

$\tan B = \frac{b}{c}$

$\because \frac{c}{b} = \frac{1}{\tan B}$

QUESTION 195

Choice D

Since $i^2 = -1$

$i^3 = i^2 \times i = -1 \times i = -i$

$\therefore i^3 + i^2 = -i + (-1) = -i - 1$

QUESTION 196

Choice C

$\sqrt{-3} + \sqrt{-9} + \sqrt{-16}$

$= \sqrt{3}i + \sqrt{9}i + \sqrt{16}i$

$= \sqrt{3}i + 3i + 4i$

$= \sqrt{3}i + 7i$

QUESTION 197

Choice B

By quadratic formula

$m^2 + 6m + 10 = 0$

$$\therefore m = \frac{-6 \pm \sqrt{6^2 - 4\times1\times10}}{2\times1} = \frac{-6 \pm \sqrt{-4}}{2}$$

$$= \frac{-6 \pm 2i}{2} = -3+i \text{ or } -3-i$$

QUESTION 198

Choice A

Since $1776 = 4 \times 444$

$i^{1776} = (i^4)^{444} = (1)^{444}$

$\therefore 704 \times i^{1776} = 704$

QUESTION 199

Choice A

$$\frac{5+7i}{6-3i} = \frac{(5+7i)(6+3i)}{(6-3i)(6+3i)} = \frac{30+15i+42i+21i^2}{45}$$

$$= \frac{30+57i-21}{45} = \frac{9+57i}{45}$$

QUESTION 200

Choice D

$$\frac{1}{2+5i} - \frac{4+3i}{3-i} = \frac{1\times(3-i)-(4+3i)(2+5i)}{(2+5i)(3-i)}$$

$$= \frac{3-i-(8+20i+6i+15i^2)}{(2+5i)(3-i)}$$

$$= \frac{3-i-(8+26i-15)}{(2+5i)(3-i)} = \frac{10-27i}{(2+5i)(3-i)}$$

Digital SAT MATH

PRACTICE TEST FOR FULL MARKS

Final Practice Test1

Module 1

QUESTION 1

Choice A

$4x+1 = -ax - 4$

$(4+a)x = -5$

when $a = -4$, this equation has no solution.

when $a \neq -4$, this equation has one solution.

QUESTION 2

Answer : $\frac{5}{2}$

Slope of $2x+5y=20$: $-\frac{2}{5}$

Slope of $y = kx - 3$: k

Since two lines are perpendicular

$\therefore -\frac{2}{5} \times k = -1$

$\therefore k = \frac{5}{2}$

QUESTION 3

Choice A

If one of the kids owns 16 of the comic books among three kids own a total of 96 comic books, the total number of comic books owned by the other two kids is 80 books.

$\therefore$ Average of comic books of the other two kids $\frac{80}{2} = 40$

QUESTION 4

Choice C

Let the value of miles did the bird travel in 5 hours be x

72 miles : 6 hours = x miles : 5 hours

$6x = 72 \times 5$

$\therefore x = 12 \times 5 = 60$

QUESTION 5

Choice C

Let the number of Franklin kite : $3x$ (each costing \$16)

and the number of Richard kite : $2x$ (each costing \$20)

$$\therefore \text{Average} = \frac{3x \times 16 + 2x \times 20}{3x + 2x} = \frac{48+40}{5} = \frac{88}{5} = 17.6$$

QUESTION 6

Choice D

1 minute : 30×15 meters = t minute : $30 \times 15 \times t$ meters

$\therefore$ Total distance d, lined by the cones as a function of t , the time in minutes = $450t$

QUESTION 7

Answer : 1

$16x^2 - 8x - 3 = 0$

$$\begin{array}{ccc} 4x & \diagdown\!\!\!\!\diagup & +1 \\ 4x & \diagup\!\!\!\!\diagdown & -3 \end{array}$$

$(4x+1)(4x-3) = 0$

$x = -\frac{1}{4}$ or $x = \frac{3}{4}$

$\therefore q = \frac{3}{4}$ or $r = -\frac{1}{4}$

$\therefore q - r = \frac{3}{4} - (-\frac{1}{4}) = 1$

QUESTION 8

Choice A

$$\sqrt{0.05} \times \sqrt{15} = \sqrt{\frac{5}{100}} \times \sqrt{15} = \sqrt{\frac{1}{20}} \times \sqrt{15} = \sqrt{\frac{15}{20}}$$

$$= \sqrt{\frac{3}{4}} = \frac{\sqrt{3}}{2}$$

QUESTION 9

Choice D

Circumference of circle $B = 2\pi B$

Circumference of circle $A = 2\pi A$

$2\pi B = 4\times(2\pi A)$

$\therefore B = 4A$

Since $A = 3$ $\quad\therefore B = 12$

$\therefore 2B - 2A = 24 - 6 = 18$

QUESTION 10

Choice A

Condition : $\dfrac{ab}{a-b} < 0$

It is efficient to find an answer to such a problem by substituting an appropriate number using a given multiple choice.

For example, when $a < b < 0$ like choice A, it is possible to check whether the inequality is satisfied by substituting appropriate numbers like $a = -2$, $b = -1$ that meet the conditions.

QUESTION 11

Choice D

Because a large cube is cut to form a sphere, it has the same ratio of properties of the same substance.

Weight of sphere : Volume of sphere = Weight of cube : Volume of cube

$52g : \frac{4}{3}\pi\times(\frac{5}{2})^3\ cm^3 = x\ g : 5\times5\times5\ cm^3$

From this equation, $x = \dfrac{52}{(\frac{4}{3}\pi)}\times 2^3 = 99.3126$

$\therefore$ The weight of meterial removed $99.3126-52=47.312 \fallingdotseq 47\ g$

QUESTION 12

Choice B

A equation for line G has to have a slope of 5 and negative y-intercept like choice B, $g(x)=5x-5\sqrt{2}$

QUESTION 13

Choice D

Let $x+2=A$

$\frac{1}{3A}=\frac{A}{48}$

$\therefore 3A^2=48$

$\therefore A=4$ or -4

QUESTION 14

Choice C

Let $RQ=x$.

Area of parallelogram $QRST=120=QT\times RU=24\times RU \quad \therefore RU=5$

Since $\angle RST=\angle RQT=30^\circ$

$\frac{5}{QR}=\sin 30^\circ \quad \therefore QR=10$

QUESTION 15

Choice C

A movie theater charges $11.50 for an Adult ticket and $9.75 for a Child ticket. But

Since both tickets were discounted by 80%, and 7 Adult tickets and x Child tickets were sold, respectively, the following equation can be established.

$11.5\times0.8\times7+9.75\times0.8\times x=95.6$ (total cost by buying tickets)

$\therefore\ x=4$

QUESTION 16

Choice A

The first term of a sequence is the number n, and each term thereafter is 5 greater than the term before. Let the first nine terms of this sequence

$n, n+5, n+10, n+15, n+20, n+25, n+30, n+35, n+40.$

The average of these nine terms of this sequence

$$=\frac{9n+(5+10+15+20+25+30+35+40+45)}{9}=n+20$$

QUESTION 17

Choice B

$2y+3 : 5 = 5y-4 : 7$

$14y+21 = 25y-20$

$11y = 41$

$\therefore\ y=\frac{41}{11}$

QUESTION 18

Choice D

Since the circumference of a right cylinder is half its height, if the radius of the cylinder is x. the volume of this cylinder $=(\pi x^2)\times(4\pi x)=4\pi x^3$

QUESTION 19

Choice C

The sum of the areas of the three rectangles $=2\times1+2^2\times1+2^2\times1=14$

QUESTION 20

Choice A

Line F is represented by the equation $y=x+1$. Line G is represented by the equation $y=px+q$. Line F and Line G always intersects where $x=1$.

When $x=1$, $y=2$ by the equation $y=x+1$.

$\therefore 2=p\times1+q \qquad \therefore p=-q+2$

QUESTION 21

Choice B

Since the quadrilateral $ABCD$ is a paralellogram,

$\angle ABD = \angle a = 30^\circ$

on $\triangle$BDC, $70^\circ + 30^\circ + b = 180^\circ$ $\therefore b = 80^\circ$

$\therefore b - a = 80 - 30 = 50$

QUESTION 22

Choice A

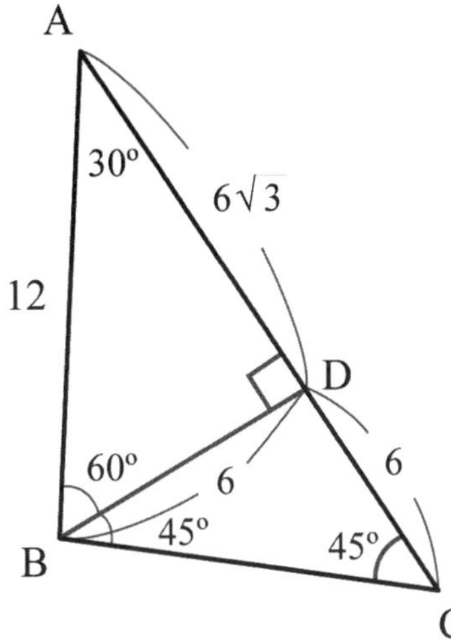

By special right angled triangle ratios, You can get the figures as shown in the picture above. $\therefore$ Area of $\triangle ABC = \frac{1}{2} \times BD \times AC = \frac{1}{2} \times 6 \times (6\sqrt{3} + 6) \fallingdotseq 49.2$

Final Practice Test1

Module 2

QUESTION 1

Choice A

As the formula of *Average speed* =,

$$Average\ speed = \frac{230 + 345 + 598}{\frac{230}{20} + \frac{345}{30} + \frac{598}{40}} = 35.545 \fallingdotseq 36$$

QUESTION 2

Choice D

If the number of elements, x doubles every hour, after 24 hours, the number will be $2^{24}x$.

Increase amount = Final amount − origin amount

$$= 2^{24} \times x - x$$

QUESTION 3

Choice A

$$5^{\frac{1}{3}} - 5^{\frac{4}{3}} = 5^{\frac{1}{3}}(1 - 5^{\frac{3}{3}}) = 5^{\frac{1}{3}}(1-5) = -4 \times 5^{\frac{1}{3}}$$

QUESTION 4

Choice B

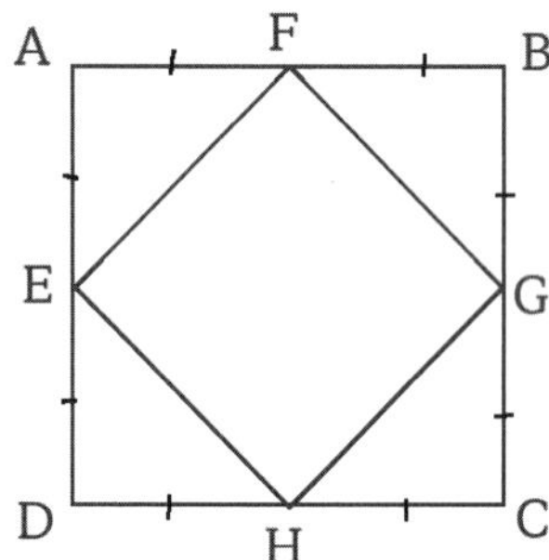

The midpoints of the sides of a square *ABCD* are connected to form a new inscribed square *EFGH*. So the area of the original square *ABCD* is two times greater than inscribed square *EFGH*.

QUESTION 5

Choice D

L x M x N $3y$ O $4y$ P

$$\begin{cases} 2x+7y=28 \\ x+3y=13 \end{cases} \Rightarrow \begin{cases} 2x+7y=28 \\ 2x+6y=26 \end{cases} \qquad \therefore y=2 \text{ and } x=7$$

$$\therefore LO : MP = \frac{2x+3y}{x+7y} = \frac{14+6}{7+14} = \frac{20}{21}$$

QUESTION 6

Choice D

The proportion of the total New York City workforce are unemployed women

$= \frac{3500}{45,500} = \frac{1}{13}$

QUESTION 7

Choice C

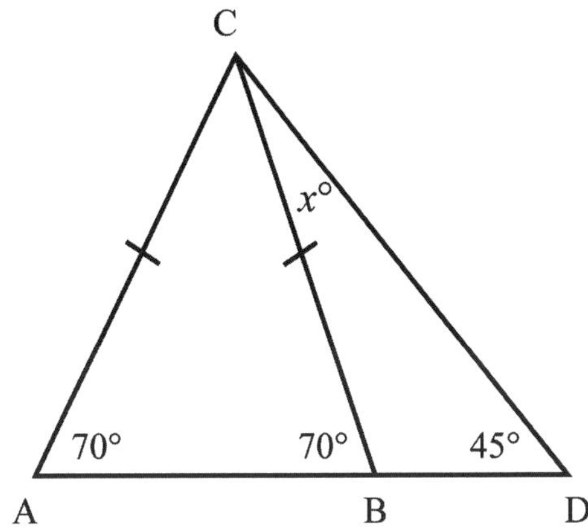

Since $\triangle ABC$ is an isosceles triangle, $\angle CAB = \angle CBA = 70^\circ$.

And as the sum of the two interior angles of the triangle is equal to the value of the remaining exterior angle, on $\triangle BCD$, $x^\circ + 45^\circ = 70^\circ$

$\therefore x = 25^\circ$

QUESTION 8

Choice C

The conditions presented in the problem can be summarized in the table as follows.

	Adult
1st best selling book	$2200 \times \frac{2}{5} \times 0.4 = 352$
2nd best selling book	x
Total number of best selling book	$2200 \times \frac{3}{8} = 825$

$\therefore\ 2200 \times \frac{2}{5} \times 0.4 + x = 2200 \times \frac{3}{8}$

$\therefore\ 352 + x = 825$

$\therefore\ x = 473$

QUESTION 9

Choice B

If the diameter of a circle is increased by 80%, the radius of a circle is also increased by 80%.

So, the area of a circle is convert from πr^2 to $\pi(1.8r)^2$.

$$\text{Increase percents} = \frac{\pi \times (1.8r)^2 - \pi r^2}{\pi r^2} \times 100\,(\%)$$

$$= \frac{1.8^2 - 1}{1} \times 100\,(\%)$$

$$= 224\,(\%)$$

QUESTION 10

Choice C

The origin split cost $= y$

The origin total cost $= xy$

The final total number of guests $= x - z$

$\therefore$ The final split cost $= \dfrac{xy}{x-z}$

$$\therefore \text{Increase percents of split cost} = \frac{\frac{xy}{x-z} - y}{y} \times 100\,(\%)$$

$$= \frac{\frac{xy - y(x-z)}{x-z)}}{y} \times 100\,(\%)$$

$$= \frac{100yz}{y(x-z)}$$

QUESTION 11

Choice D

In the figure,

Radius of circle B $= \dfrac{3}{4} \times$ Radius of circle A

$$(x+2) = \frac{3}{4}(3+x)$$

$$4x + 8 = 9 + 3x$$

$$\therefore x = 1$$

Since radius of $B = 2 + x = 3$

$\therefore$ Area of circle $B = \pi \times 3^2 = 9\pi$

QUESTION 12

Choice B

Let the number of several exam scores except one make-up exam(92 scores) was given, x

Then total sum of scores of all exam scores : $80x + 92$

Since the new average was 84 including one make-up exam(92 scores)

$\therefore \dfrac{80x + 92}{x + 1} = 84$

$\therefore x = 2$

$\therefore$ Let the number of several exam scores including one make-up exam $= x + 1 = 3$

QUESTION 13

Choice A

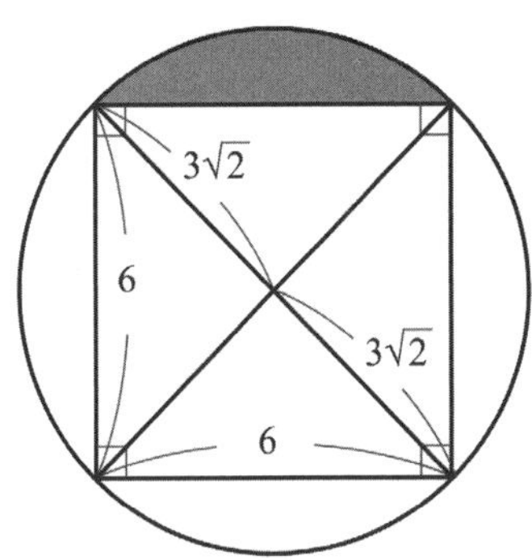

Arc length of shaded region $= 3\sqrt{2} \times (\dfrac{\pi}{2}) = \dfrac{3\sqrt{2}\,\pi}{2}$

$\therefore$ Perimeter of shaded region $= 6 + \dfrac{3\sqrt{2}}{2}\pi$

QUESTION 14

Choice C

Slope of the line $l = \dfrac{(b+c) - b}{(a + b) - a} = \dfrac{c}{b}$

Since $(13, 10)$ is a point on the line l,

An equation of line l is

$y-10=\frac{c}{5}(x-13)$

And since the x-intercept of line l is -7,

$0-10=\frac{c}{5}(-7-13)$

$\therefore c=2.5$

QUESTION 15

Choice D

Since the present age of Eric is $22+x$ years old and Shelly is $24-y$ years old, Their ages 4 years ago were $18+x$ years old and $20-y$ years old.

$\therefore$ The average of their ages 4 years ago $=\frac{(18+x)+(20-y)}{2}=\frac{38+x-y}{2}$

QUESTION 16

Choice A

The conditions presented in the problem can be summarized in the diagram as follows

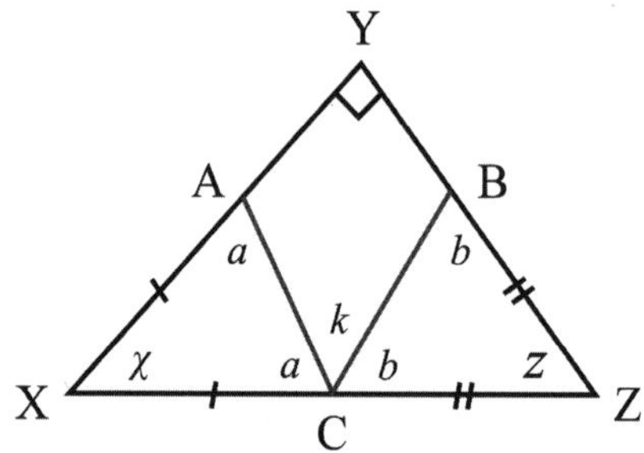

Since $2a+x=180°$ and $2b+z=180°$

From the sum of two equations, $2a+2b+x+z=360°$

And as $x+y=90°$ $\therefore$ $2a+2b=270°$

$\therefore a+b=135°$

On XCY, $a+k+b=180°$

$\therefore$ $\angle ACB=k=45°$

QUESTION 17

Choice A

In this problem, It is efficient to find an answer to such a problem by substituting an appropriate number using a given multiple choice.

For example, if the center of the circle is $(15, 18)$ like Choice A,

The distance between $(15, 18)$ and point A :

$r = \sqrt{(15-3)^2+(18-13)^2} = \sqrt{12^2+5^2} = 13$

The distance between $(15, 18)$ and point B :

$r = \sqrt{(15-15)^2+(18-5)^2} = \sqrt{0^2+13^2} = 13$

QUESTION 18

Choice D

$y = h(x)$

↓ Horizontal stretching factor : 2, vertical stretching factor : $\frac{1}{2}$

$y = \frac{1}{2}h(\frac{1}{2}x)$

↓ reflection to y axis

$y = \frac{1}{2}h(-\frac{1}{2}x)$

QUESTION 19

Choice D

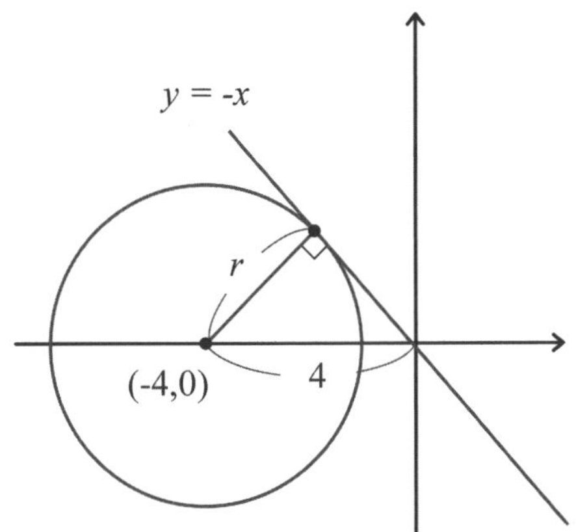

The conditions presented in the problem can be summarized in the diagram as follows

Since $r = 2\sqrt{2}$, the circumference of the circle $= 2\pi r = 2\pi \times (2\sqrt{2}) = 4\pi\sqrt{2}$

QUESTION 20

Choice D

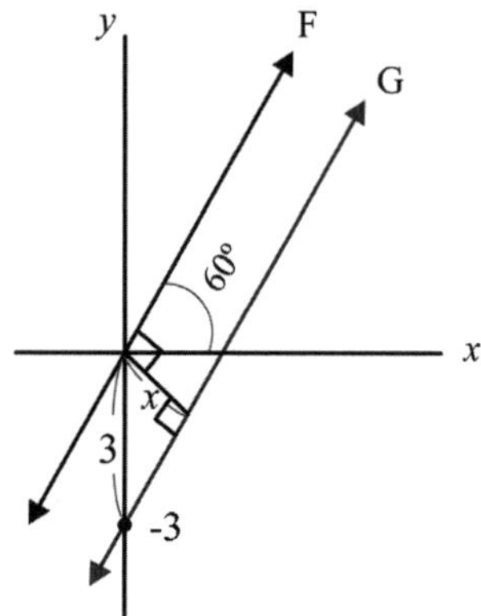

By special right angled triangle ratios, the shortest distance between two parallel lines F and G, x is $\frac{3}{2}$.

QUESTION 21

Choice D

$(2+3i)(3+ai) = a \in$ Real number set

$6 + 2ai + 9i + 3ai^2 = 0$

$2a + 9 = 0$

$\therefore a = -\frac{9}{2}$

QUESTION 22

Choice C

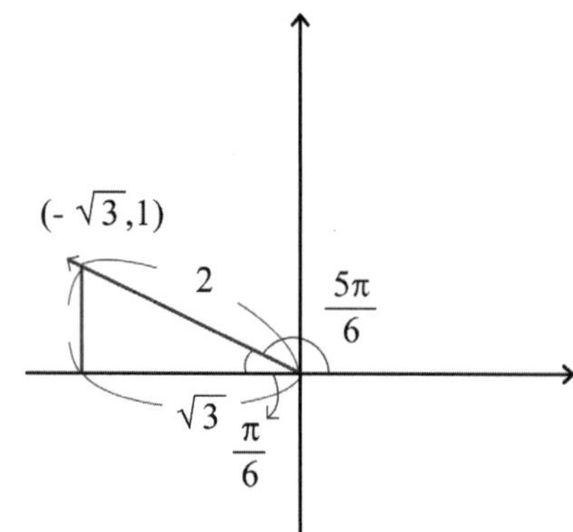

By the definition of trigonometric ratio on xy plane

$\sin\left(\frac{5\pi}{6}\right) = \frac{y}{r} = \frac{1}{2}$

Final Practice Test2

Module 1

QUESTION 1

Choice B

Let the number of David's nickles $= D$, the number of Tom's nickles $= T$, and the number of John's nickles $= J$

$D = 2T$

$T = J + 15$

Since $J = 6$ (given), $T = 21$, $D = 42$

$\therefore 42 \times \$0.05 = 2.1$

QUESTION 2

Choice C

Since 10 pencils $\leq$ 1 case $\leq$ 14 pencils

Since 50 pencils $\leq$ 5 case $\leq$ 70 pencils

QUESTION 3

Choice B

$\frac{13\pi}{12} = \frac{13}{12} \times 180^\circ = 195^\circ$

QUESTION 4

Choice C

Let the central angles of two piece of cakes, $3x$ and $5x$.

then $3x + 5x = 360^\circ$

$x = 45^\circ$

$\therefore 3x = 135^\circ$

QUESTION 5

Choice C

Since $\sqrt{72} = \sqrt{6\times6\times2} = 6\sqrt{2}$

$\therefore 6\sqrt{2} + 6\sqrt{2} = 12\sqrt{2}$

$\therefore m = 12$

QUESTION 6

Answer : $\frac{1}{3}$

The slope of line l with equation $2x + 3y = 6$ is $-\frac{2}{3}$. Since line $y = ax + b$ passes through the point $(0, 1)$ and is parallel to tIf the equation of line l,

$y-1 = -\frac{2}{3}(x-0)$

$\therefore y = -\frac{2}{3}x + 1$

$\therefore a = -\frac{2}{3},\ b = 1$

$\therefore a + b = \frac{1}{3}$

QUESTION 7

Choice B

The slope of $y = \frac{1}{12} - x$ is -1, and y intercept is $\frac{1}{12}$. So the line has a negative slope and a positive y-intercept.

QUESTION 8

Choice B

The discriminant of a quadratic equation $3x^2 + kx + 3 = 0$ is $k^2 - 4\times3\times3$.

Since this quadratic equation has only one solution,

$k^2 - 36 = 0$

$\therefore k = 6$ or -6

QUESTION 9

Choice B

Given $g(x) = 8x - 5$

$g(g(x)) = g(8x-5) = 8(8x-5) - 5 = 64x - 45$

QUESTION 10

Choice B

$30 = 1.78 \times \text{voltage} + 1.5$

$\therefore$ voltage $= 16.011 \fallingdotseq 16$

QUESTION 11

Choice C

Check filled by a pump at a rate $= 200\ ft^3/\text{min}$.

And let the draining out of rate $= x$

$\therefore (200 - x) \times 200 = 35000$

$200 - x = 175$

$\therefore x = 25$

QUESTION 12

Choice C

Since $\angle OAC = \angle ODB$ and $OD = DB$, $\triangle ODB$ is a equilateral triangle.

$\therefore DB = 6$

QUESTION 13

Choice A

$\dfrac{8^x}{2^y} = \dfrac{(2^3)^x}{2^y} = 2^{3x-y} = 2^{12}$

QUESTION 14

Choice B

If Te two airlines with the highest flight delays from the dot plot is removed, the mean and range will decrease only. but median is the same.

QUESTION 15

Answer : 4.8

By Pythagorean theorem,

$d^2 = 3.8^2 + 2.9^2$

$\therefore d = 4.7801 \cdots \fallingdotseq 4.8$

QUESTION 16

Choice C

$$\text{Volume of square pyramid} = \frac{1}{3} \times \text{Base area} \times \text{Height}$$

$$= \frac{1}{3} \times 3^2 \times 3 = 9$$

$\therefore$ The time to take to print one pyramid = Total volume ÷ Speed

$$= \frac{9}{0.025} = 360 \text{ seconds} = 6 \text{ minutes}$$

QUESTION 17

Choice A

When $x = 0$, $g(0) = 2 \times (-5) \times (-3) = 30$

$h(0) = 2 \times 5 \times 3 = 30$

So two functions have the same y-intercept.

QUESTION 18

Choice A

Let the value of House 8 $= x$

$$\text{Then the average of All eight houses} = \frac{180 + 200 + 225 + 250 + 252 + 255 + 256 + x}{8}$$

$$= \frac{1618 + x}{8} \geq 251$$

$\therefore 1618 + x \geq 251 \times 8$

$\therefore x \geq 390$ (in thousands)

QUESTION 19

Choice D

The area of shaded sector $= \dfrac{1}{2}r^2(\pi - 0.6) = 6$

$r^2 = \dfrac{12}{\pi - 0.6} \qquad \therefore r = \sqrt{\dfrac{12}{\pi - 0.6}}$

QUESTION 20

Choice C

On $v(t) = 0.41 \times 0.8^{\frac{2t}{5}}$

when $t = \dfrac{5}{2}$ $v(\dfrac{5}{2}) = 0.41 \times 0.8^{\frac{2}{5} \times \frac{5}{2}} = = 0.41 \times (1 - \dfrac{20}{100})^1$

This means it is exponential because the voltage decreases by 20% every 2.5 minutes.

QUESTION 21

Choice B

Assuming that the cylinder is unfolded, the following figure is obtained, and the value obtained is a diagonal line of a rectangle.

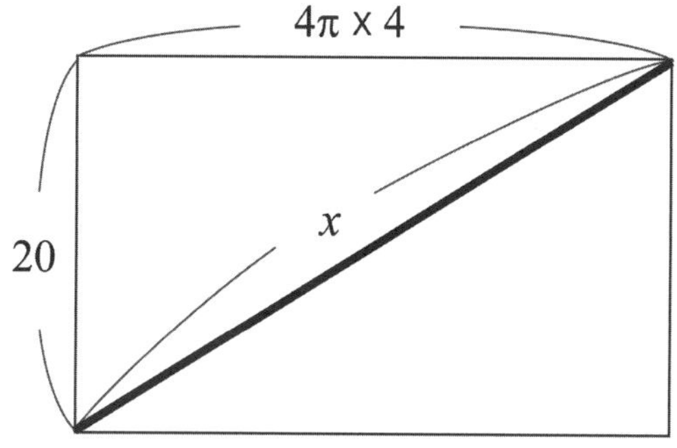

$x = \sqrt{(16\pi)^2 + 20^2} = 54.098\cdots \fallingdotseq 54$ ft

QUESTION 22

Choice C

The conditions presented in the problem can be summarized in the diagram as follows

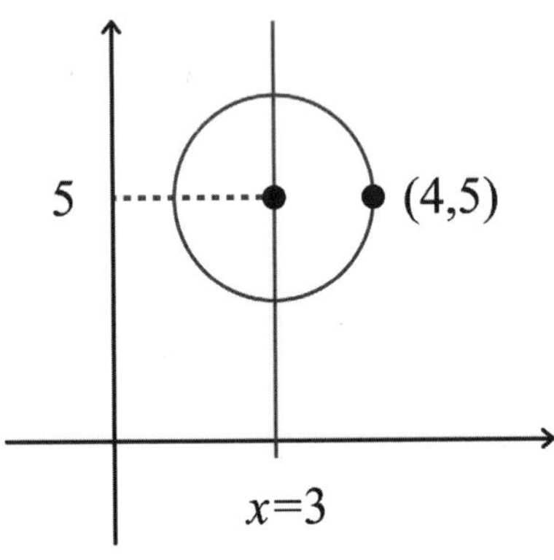

$\therefore r = 1$

Final Practice Test2

Module 2

QUESTION 1

Choice B

Let the number of the scientists $5x$ and engineers x. And assuming in the lab, After 75 new team members were hired in the numbers of engineers $2y$ and scientist y, since the ratio of scientists to engineers to approximately $2:3$ you can set the following equation and the ratio equation.

$2y + y = 75 \quad \therefore y = 25$

$5x + y : x + 2y = 2 : 3$

$\therefore 5x + 25 : x + 50 = 2 : 3$

$\therefore x = 9.615 \cdots \fallingdotseq 10$

question 2

Choice A

Since the model equation $P = 175 + \frac{11}{2}t$ gives the population, P, in thousands, with respect to time, t (provided that 2010 is when $t = 0$).

175 means the population of Sub−Saharan African was 175 thousand In 2010.

question 3

Choice D

$$\begin{cases} 2x + 5y = 20 \\ 5kx - 5y = 15a \Leftarrow y = kx - 3a \end{cases}$$

From adding two equations

$(2+5k)x = 20+15a$

When $2+5k = 0$ and $20 + 15a = 0$, two lines are the same lines.

$\therefore k = -\dfrac{2}{5}$ and $a = -\dfrac{4}{3}$

$\therefore ka = \dfrac{8}{15}$

question 4

Choice C

Since $P_A = P_B$

$100 \times e^{0.02t} = 2 \times 2000 \times e^{-0.01t}$

$e^{0.02t} = 40 \times e^{-0.01t}$

By Graphing calculator,

$\therefore t = 122.96 \cdots \fallingdotseq 123$

question 5

Choice B

Let the green tomatoes $4x$ and red tomatoes $3x$.

The conditions presented in the problem can be summarized in the ratio equation as follows.

$4x - 5 : 3x - 5 = 3 : 2$

$9x - 15 = 8x - 10$

$\therefore x = 5$

$\therefore$ The number of the origin red tomatoes $= 3x = 15$

question 6

Answer : 113

The common difference of two consecutive terms is 7.

$\therefore (2a-1) - a = 7$

$\therefore a = 8$ $\therefore$ 16th term $= 8 + 7 \times 15 = 113$

question 7

Choice D

$p^3 = 1 \times p^3 = p \times p^2$

Since p ia a prime factor, p has four factors ; $1, p, p^2, p^3$

question 8

Answer: 16

$(a-4)(b+6)=0$

$a=4$ or $b=-6$

The smallest possible value of $a^2+b^2=16$ when $a=4$ and $b=0$.

question 9

Choice D

Given that $f(x)=g(x)$

$ax^2 = bx^4$

$x^2(bx^2-a)=0$

$\therefore x^2=0$ or $x^2=\frac{a}{b}$

$\therefore x=0$ or $x=\sqrt{\frac{a}{b}}$ or $x=-\sqrt{\frac{a}{b}}$

question 10

Choice D

If the water filling speed is the same, the wide width is slow to fill, and the narrow width is fast to fill. The cross-section shown in the problem is wide at the bottom, narrow at the midpoint, and wide at the last top. Therefore, the rate of water filling is slow at first, then fastest at the midpoint, and slow again at the end.

question 11

Choice D

As given that $N(t) = k \times 2^{-at}$,

when $t=1$, $N=16=128\times 2^{-a}$

$\therefore 2^{-a}=\frac{1}{8}=2^{-3}$

$\therefore a=3$

question 12

Choice B

Since the value of y increased by 12 is directly proportional to the value of x decreased by 6, $y+12=k(x-6)$

If $y=2$ when $x=8$, $14=k\times 2$ $\quad\therefore k=7$

when $y=16$, $28=7\times(x-6)$

$\therefore x-6=4$

$\therefore x=10$

question 13

Choice B

By special right angled triangle($30^\circ, 60^\circ, 90^\circ$) ratios and the conditions presented in the problem can be summarized in the diagram as follows.

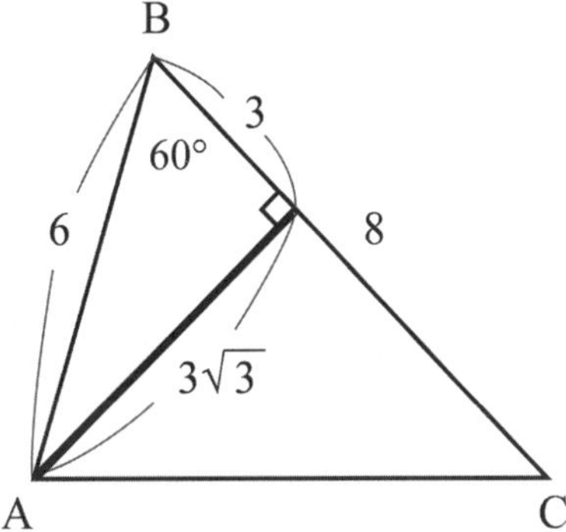

$\therefore$ Area of $\triangle ABC=\frac{1}{2}\times 8\times 3\sqrt{3}=12\sqrt{3}$

question 14

Choice D

The slope between $(0,1)$ and $(b,0)$ $=\frac{0-1}{b-0}>-\frac{1}{2}$

$-\frac{1}{b}>-\frac{1}{2}$

$$\frac{1}{b} < \frac{1}{2}$$
$$\therefore b > 2$$

question 15

Answer: $2 < x < 3$

$x + c = 2x - 1$

$x = c + 1$

Since $1 < c < 2$

$\therefore 2 < c + 1 < 3$

$\therefore 2 < x < 3$

question 16

Choice D

Since VitaDrink contains 30 percent($0.3v$) concentrated nutrients by volume. EnergyPlus contains 40 percent($0.4e$) concentrated nutrients by volume and the volume in a mixture of v gallons of VitaDrink, e gallons of EnergyPlus, and w gallons of water are $v + e + w$.

$$\therefore \frac{0.3v + 0.4e}{v + e + w} \times 100 = \frac{30v + 40e}{v + e + w}$$

question 17

Choice B

If any one number x among the seven numbers whose original average value is 12, if the average is 15, then the number x changed is bound to be less than 6. so

$7 \times 12 = 84$

$7 \times 15 = 105$

$\therefore 6 - x = 21$

$\therefore x = -15$

question 18

Choice D

If the values of a, b and c be positive integers. and the average (arithmetic mean) of a, b and c is 100, $a + b + c = 300$.

Since $a \geq 1$, $b \geq 1$ and $c \geq 1$, one of the three numbers cannot be 299.

question 19

Choice D

Example 1 : $M = \{1, 2, 3, 4, 5\}$ $\Rightarrow$ mean: 3, median: 3

Example 2 : $M = \{2, 3, 4, 5, 6\}$ $\Rightarrow$ mean: 4, median: 4

Example 3 : $M = \{1, 2, 3, 4, 5, 6\}$ $\Rightarrow$ mean: 3.5, median: 3.5

I. The average (arithmetic mean) of the numbers in set M equals the median. $\Rightarrow$ True.

II. The number of numbers in set M is odd. $\Rightarrow$ True since 3.5 is not including M on example 3

III. The sum of the smallest number and the largest number in set M is even. $\Rightarrow$ True.

question 20

Choice B

$4|6+2s| \le 24$

$|6+2s| \le 6$

$-6 \le 6+2s \le 6$

$-12 \le 2s \le 0$

$\therefore -6 \le s \le 0$

question 21

Answer : 202

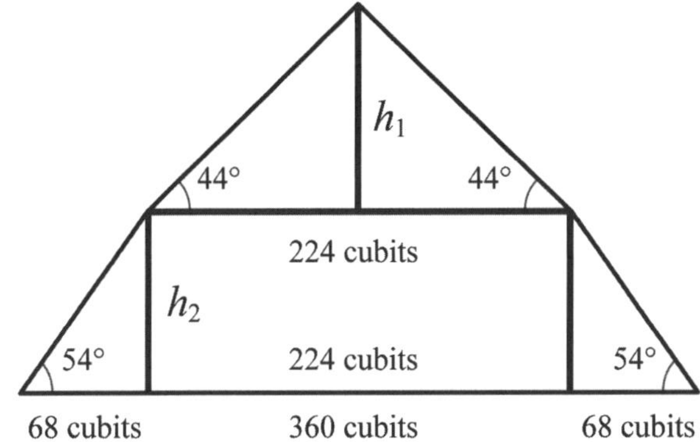

$\frac{h_1}{112} = \tan 44^\circ$, $\frac{h_2}{68} = \tan 54^\circ$

$\therefore$ The height of the pyramid to the nearest cubit $= h_1 + h_2$

$= 112 \times \tan 44^\circ + 68 \times \tan 54^\circ$

$= 201.751 \cdots \fallingdotseq 202$

question 22

Answer : 37.5

Speed of $A = \frac{r}{15}$

Speed of $B = \frac{r}{25}$

After t seconds,

$$\frac{r}{15} \times t - \frac{r}{25} \times t = r$$

$$\frac{t}{15} - \frac{t}{25} = 1$$

$$\frac{25t - 15t}{25 \times 15} = 1$$

$$10t = 25 \times 15$$

$$\therefore t = 37.5$$

How to use a calculator on The Digital SAT Math (Using by Ti-84 Plus/Ti-Nspire)

1. 426
2. 2.2635
3. 100.5491
4. 1275
5. 72.1481
6. 4.096
7. 13.57208
8. 8.78623

9~11. $x = -4$ or 1

12i) $(-0.75, -6.125)$

12ii) $(0, -5)$

12iii) $(-1, 0)$ or $(2.5, 0)$

13~14. 4.8298

15. $x = 10,\ y = 100$

16. C